H. Kirke (Harry Kirke) Swann

A concise Handbook of British Birds

H. Kirke (Harry Kirke) Swann

A concise Handbook of British Birds

ISBN/EAN: 9783743324671

Manufactured in Europe, USA, Canada, Australia, Japa

Cover: Foto ©berggeist007 / pixelio.de

Manufactured and distributed by brebook publishing software
(www.brebook.com)

H. Kirke (Harry Kirke) Swann

A concise Handbook of British Birds

A CONCISE HANDBOOK

OF

BRITISH BIRDS.

A CONCISE HANDBOOK

OF

BRITISH BIRDS

BY

H. KIRKE SWANN

Editor of "THE ORNITHOLOGIST."

London:

JOHN WHELDON & CO.,

58, GREAT QUEEN STREET W.C.

1896.

TO

JOSEPH WHITAKER, F.Z.S., M.B.O.U.

A FRIEND,

THIS BOOK IS DEDICATED BY

THE AUTHOR.

PREFACE.

ALTHOUGH the urgent need for compression has prohibited the author from presenting this work as a *complete* natural history of British Birds, yet he believes that as a handy textbook for reference it has had as yet no rivals. It has been brought up to date as far as possible, every species on the British list having been included, and nearly all described, while the records of the rarer species have been carefully collected, and in the case of the common species a tolerably complete life-history has been given. The numbering of the genera and species will also probably prove of service.

The length of a bird is (unless otherwise stated) measured from the base of bill to tip of tail, all measurements being in inches and hundredth-parts. The description first given is (unless expressly mentioned) that of a bird in breeding plumage, but the seasonal changes, if appreciable, are added. The *habitat* may be taken as meaning the region inhabited during the breeding season, but the winter range of migratory species has usually been added. If brackets enclose a describer's name they indicate that the *generic* name adopted is not that used by the describer.

The classification and nomenclature practically accord with those of the "List of British Birds" compiled by a Committee of the British Ornithologists' Union (1883), but a number of necessary alterations have been made, particularly in the matter of adopting the specific names of *first* describers as far as possible. An effort has also been made to allow specific rank to valid species only, while sub-species or races, instead of being nameless, are distinguished by sub-numbers and trinominals—after the American style. With the exception of the late Henry Seebohm, no British ornithologist appears to have openly avowed himself a trinominalist, so that I shall not escape censure for adopting the despised system, yet until some of our ornithologists can suggest some other way of allowing a name to a recognised race without giving it the rank of a species, I will adhere to trinominals.

My thanks are due to various friends who have assisted me with notes and information, much of which, however, I have been unable to make use of in this edition for want of space. I am especially indebted to Dr. John Trumbull, of Malahide, Co. Dublin, for sending me annotated copies of the late A. G. More's "List of Irish Birds" (1890), and Mr. Ussher's "Report on the Breeding Range of Irish Birds" (1894), the interleaved MS. notes bringing the bibliography down to the present time. I have also obtained much help from Mr. Howard Saunders' "Manual of British Birds" (1889), Mr. J. E. Harting's "Handbook of British Birds" (1872), Dr. Sharpe's "Handbook to the Birds of Great Britain" (1894—6), Mr. Dresser's "Birds of Europe" (1871—81), Seebohm's "British Birds" (1883—5), Prof. Collett's "Bird Life in Arctic Norway" (1894), the 4th edition of Yarrell, and the 2nd edition of the A. O. U. Check-list of North American Birds, also the "Zoologist," "Ibis," "Field," etc.

H. K. S.

London, June 6th, 1896.

LIST OF GENERA.

A CONCISE HANDBOOK

OF

BRITISH BIRDS.

ORDER PASSERES.

Family Turdidæ.

Sub-Family Turdinæ.

GENUS I. TURDUS, *Linnæus (1766)*.

Bill of medium length, straight, tip of upper mandible a little decurved and slightly notched. Wing with first quill small, 3rd usually longest. Tail moderately long. Tarsus rather long; claws large, curved.

(NOTE.—In all species of *Turdinæ* young are spotted both above and below; this plumage, however, being lost in first autumn moult.)

1. Turdus viscivorus, Linn. MISTLE-THRUSH.

Habitat. Europe, north to lat. 67° in Norway; also extreme north of Africa, Asia Minor, and Asia, east to Lake Baikal, and south to Himalayas. In winter migrates from most northern regions.

Adult male: upper plumage greyish brown; wings darker; rump with a yellowish tinge; lower parts whitish, boldly marked with large spots of dusky-brown; throat with a yellowish tinge, and marked with smaller arrow-shaped spots; under wing-coverts white; bill dark brown, yellowish at lower base; iris hazel; tarsi yellowish-brown; claws dusky. Length 11·00. Female scarcely differs. Young, after first moult, have lower parts tinged with yellowish buff, while upper wing-coverts are noticeably tipped with white.

Common throughout Great Britain, north to Hebrides. In Ireland now breeds in every county, although seemingly unknown there 100 years ago. Song loud and not possessed of much variety; usually commenced in January. On the wing may be distinguished by its large size, irregular and rapid flight, and loud harsh "churring" note. Food: mountain-ash, holly and hawthorn berries, also insects, snails, etc., upon which young are fed; rather partial to fruit. Nest: usually in forks of trees in woods or orchards; sometimes upon small branches against trunk; height varies from 10 to 30 feet; once I found it 8 feet up in a hedge; composed of small twigs, grass-stalks, lichens, paper, etc., plastered with mud and lined with much dry grass. Eggs: 3 to 5; varying from greenish to reddish-white, with spots and blotches of reddish-brown, and underlying lilac markings; size 1·30 by ·86.

2. Turdus musicus, Linn. SONG-THRUSH.

Hab. Europe (except extreme south) and Asia, east to Pacific, north to about 60° N. lat.; in Norway to 70° N. In winter reaches North Africa and Persia.

Male: above warm olive-brown, wing-coverts being margined with buff; lower parts buffish-white with dark brown streaks and spots; centre of abdomen white; under wing-coverts orange buff; bill dark brown, under mandible yellowish at base; iris hazel; tarsi pale brown.

Length about 9·00. Female a trifle smaller. Young, in first plumage, mottled above with golden-buff.

Common everywhere, north to Outer Hebrides, but rare in extreme west of Ireland. Well-known song is often commenced during last week of January, and continued until summer; also usually heard again during latter months of year. Food: berries, insects, worms, and especially snails, shells of which it breaks against a stone; also a little fruit when ripe. When alarmed it flies off nearly in a straight line, uttering a loud *cheek, cheek, cheek.* Nest, with its neat lining of decayed wood and dung, is known to almost everyone; usually in bushes or hedges, but I have found it at a height of 20 feet in trees. Eggs: 4-6; seldom found before middle of March; shell glossy, greenish-blue, blotched and spotted with reddish-brown and black; they vary much in marking, also in size, average being 1·05 by ·80. Two or three broods are produced. I have found eggs as late as August 6th.

3. Turdus iliacus, Linn. REDWING.

Hab. Siberia, West of the Yenesei, and westward (above 54° N. lat.) to Scandinavia, the Færoe Islands and Iceland. In winter south as far as N. Africa, Persia and N. India.

Male: upper plumage dark brown, with paler margins to wing-feathers; above eye a noticeable yellowish-white streak; lower parts white with a buff tinge and bold streaks of dark brown upon throat, breast and sides upper portion of latter, together with under wing-coverts, being of a conspicuous orange-red colour; bill dark brown, base of lower mandible yellowish; tarsi light brown. Length 8·75. Female differs very little.

There is no satisfactory evidence of its breeding with us. Majority leave by third week in March, although a few sometimes remain until end of April. Usually seen about the fields in small flocks in company with Fieldfares, etc., but also frequents woodlands. Alarm note a sharp but liquid *chuck*. Largely a ground feeder, food consisting of insects, snails, etc., but also fond of holly berries.

4. Turdus pilaris, Linn. FIELDFARE.

Hab. Breeds from Siberia (west of the Lena) to Finland and Scandinavia, and southward to Bavaria and Poland. In winter south to N. Africa and India.

Male: head slate-grey, streaked with black; mantle and upper wing-coverts chestnut-brown; lower back ash-grey; tail-feathers dark brown; throat and breast brownish buff streaked with black, flanks also being boldly marked with blackish-brown; centre of abdomen and under wing-coverts pure white; bill yellow; tarsi dusky brown. In winter plumage is slightly duller, and bill brownish. Length 10·25. Female scarcely differs.

There is no actual proof that it has bred with us. Arrives late in September, leaving by middle of April; stragglers, however, sometimes remaining until May. Gregarious, frequently in company with other species; feeding upon insects, grubs, and various berries. May be readily distinguished by its blue-grey lower-back when near the ground; white belly and under-wing are also noticeable during flight. Note: a loud harsh *tzer-er, tzack, tzack*, uttered on the wing; flight rapid.

5. Turdus atrigularis, Temm. BLACK-THROATED THRUSH.

Hab. Siberia, south to the Himalayas and Turkestan. In winter south to Northern India and Persia. A rare straggler to Europe.

Male: upper plumage olive-brown, with wing-quills and tail dusky; throat and breast dull black; belly white; sides brownish; bill dusky brown, paler below; tarsi light brown. Length 9·70. In winter throat-feathers have grey margins. Female: feathers of throat and breast have blackish centres only and under parts are brownish-white. Young males resemble female.

Two occurrences are on record, *i.e.*, a young male shot in Sussex in 1868 and another example shot near Perth, and recorded in the "Ibis" for October, 1889, by Lieut.-Col. H. M. Drummond Hay.

6. Turdus varius, Pall. WHITE'S THRUSH.

Hab. Eastern Siberia, North China, and Japan. In winter south as far as Philippines and casually westward to Europe.

Adult bird has feathers of upper parts brownish-buff tipped with black; under parts buffish-white, marked with crescent-shaped spots of black; tail-feathers 14 in number and tipped with white. Length fully 11·50.

Earliest known occurrence was in Hampshire (1828). Since then has occurred in most counties in east and south of England, also in Berwickshire and three times in Ireland: in counties Cork, Longford and Mayo.

7. Turdus merula, Linn. BLACKBIRD.

Hab. Europe, north and east to South Norway and Central Russia. In winter migrates from most northern regions.

Male: plumage entirely black and glossy; tail rather long; bill and eyelids yellow; tarsi brownish-black. Length 10·50. Female: upper plumage blackish-brown; throat buffish-white, spotted with reddish-brown; breast reddish-brown; belly and flanks slate-black: bill orange-brown; tarsi dark brown.

A common resident. Song is frequently commenced during last week of January, while (like that of song-thrush) it is usually heard again during autumn. The song is simpler than latter bird's, yet more mellow and rich in tone; one would think it scarcely so loud, yet it may be heard at a great distance in the still morning air. A sad destroyer of fruit; I have known two or three to strip a cherry-tree before breakfast-time. Usually, however, a ground-feeder, subsisting upon worms, insects, snails, etc.; in winter gregarious, in company with other thrushes, and feeding largely on berries. When disturbed, very energetic alarm-note is a screaming *kit*, *kit*, *kit*, *kit*, continued until well on the wing. Nest: like that of Song-Thrush but more bulky and lined with grasses; placed at no great height in hedge-rows and bushes; sometimes in a steep bank. Eggs: 4 to 6; pale greenish-blue spotted with reddish-brown, but markings are subject to variation; size 1·16 by ·85. Two or three broods are produced, first eggs being laid about end of March.

8. Turdus torquatus, Linn. RING-OUZEL.

Hab. Europe, from Ireland to Ural Mountains in north, and southward to most of mountain ranges. In winter south to North Africa.

Male: feathers of upper parts blackish-brown, margined with grey on wings; under parts similar but with a broad crescentic gorget of pure white, under wing-coverts also being mottled with white; bill brownish-yellow, dusky at tip; iris hazel; tarsi dusky. Length 10·50. Female: browner, and with a narrow and less distinct gorget.

Arrives in April, majority departing in October; occasionally, however, remains through winter in south-west of England, and also in Ireland. Breeds from Derbyshire to the north of Scotland, also in Wales, Devon and Cornwall; in Ireland breeds sparingly in all the mountainous districts. Nest: usually in the steep bank of a mountain stream or watercourse, sometimes in low bushes; similar to Blackbird's, and constructed of moss, roots, small twigs, etc., plastered with mud and lined with dry grass. Eggs: 4 or 5; greenish-blue, spotted and blotched with reddish-brown, markings being usually more distinct than in eggs of Blackbird; size 1·15 by ·85.

GENUS II. MONTICOLA, *F. Boie (1822).*

Combines part of the characteristics of *Turdus* and *Saxicola*; differs from former in having tail short and even; wings moderate, and bill and feet rather stout.

9. Monticola saxatilis (Linn.). ROCK-THRUSH.

Hab. Southern Europe and temperate Asia, eastward to North China. In winter south to Africa and N.W. India.

Male: head, nape, lower part of back, and throat, slate-blue; upper part of back dusky-blue; centre of back white; wing-feathers dusky-brown; middle tail-feathers brown, remainder chestnut; under plumage chestnut; bill black; tarsi brownish. Length 7·30. Female: mottled with brown above, and also on orange-brown lower parts; tail-feathers, chestnut; throat, white.

One occurrence only is known; an example shot in Hertfordshire (1843).

GENUS III. SAXICOLA, *Bechstein (1802)*

Wing with 3rd or 4th quill longest. Tarsus long; hind claw moderate, stout, curved.

After autumn moult most of the feathers of wings and upper plumage show broad buff margins.

10. Saxicola œnanthe (Linn.). WHEATEAR.

Hab. Whole Palæarctic Region. In winter south to Equator, also casually on Atlantic coast of America to Bermudas.

Male: forehead and stripe above eye white; ear-coverts and streak below eye black; crown, nape, and mantle, light grey; wing-feathers black; lower part of back white; tail-feathers white, broadly tipped with black, except two central ones, which have only a little white at root; lower plumage white, with a creamy tinge on throat and breast; under wing-coverts with grey mottlings; bill and tarsi black; iris hazel. Length about 6·00; wing 3·70. Female differs in having upper plumage brown with no black on head; lower plumage brownish-white.

Generally distributed, arriving in March and leaving in September. Most abundant everywhere during migration, although fair numbers remain to breed, wherever there are open hills, downs or upland pastures. I have found nest usually in unfinished rabbit-burrows; it is also placed in crevices in rough walls or under rocks and clods of earth; loosely constructed of roots, and dry grass, lined with rabbit's fur, hair and feathers. Eggs: 5 or 6 sometimes 7; surface smooth, very pale blue, usually spotless but occasionally with a few brown specks; size ·82 by ·62. Food: insects, grubs, etc., it will take flies on the wing like a Stonechat. When disturbed, flies a short distance and settles again with a few jerks of its tail. Usual note a sharp *chack, chack.* Birds which breed in Greenland, and are observed in our islands on spring migration, belong to a larger race than those which nest here, although their claim to sub-specific rank is not recognised.

11. Saxicola isabellina, Ruppell. Isabelline Wheatear.

Hab. Resident in N.E. Africa, from Abyssinia to Egypt; also in Palestine. Breeds as well in S.E. Russia and temperate Asia, migrating southward in winter.

The only recorded occurrence is that of a female shot in Cumberland (1887), and identified by Mr. Saunders.

12. Saxicola stapazina (Vieill.). Black-throated Wheatear.

Hab. Western Europe, north to the Loire in France, south to extreme North-west Africa.

Male: forehead white; crown, nape and mantle pale reddish-buff, darker after summer moult; lower back white; wings (both above and below) black; two middle tail-feathers black; remainder white, slightly margined with black; throat and cheeks black; under parts white tinged with buff; bill and feet black. Length 5·70; wing 3·60. Female: has mantle buffish-brown; wings dark-brown, and throat only mottled with black.

This species also has only occurred once, namely, in 1875, when an adult male was shot in Lancashire.

13. Saxicola deserti, Temm. Desert Wheatear.

Hab. North Africa, from Morocco to Egypt; also from Arabia and Palestine to Central Asia. Migrating southward in winter. A rare straggler to Europe.

Male: upper plumage sandy-buff; scapulars and under wing-coverts mostly black; secondary quills brown, with lighter margins; primaries black, with pale margins to inner webs; upper tail-coverts white; tail black, except just at base; cheeks and throat deep black, bordered by a white line above eye; lower plumage white, tinged with buff; bill and feet black. Length 6·00. Female: duller, greyer and lacks black on throat and wings.

Within the last fifteen years three examples have been obtained, viz., a male shot in Clackmannanshire (1880); a female in Yorkshire (1885); and a third example near Arbroath (1887).

Genus IV. PRATINCOLA, *Koch (1816)*.

Differs from *Saxicola* in having bill *shorter*, weaker, and wider at base; bristles at the gape distinctly developed; wings and tail somewhat short.

14. Pratincola rubetra (Linn.). Whinchat.

Hab. Northern and Central Europe; northward in Scandinavia to 70° N. latitude. In winter south to Abyssinia and Northern India.

Male: cheeks and ear-coverts blackish brown, bordered below by a white streak running backward to the nape, and above by a bold white stripe over the eye; crown of head, nape and back, sandy brown, streaked with dark brown; wings dark brown, blackish on coverts, on which is a noticeable patch of white; upper tail-coverts with a reddish tinge; tail-feathers white at base, with exception of two middle ones, which are dark brown, as are also terminal halves of remainder; centre of abdomen whitish; throat, breast and flanks, pale yellowish brown; bill and feet black. Length 5·25. Female: duller; eye-stripe yellowish-white; white patch on wing-coverts not so distinct. Young resemble female, but have upper feathers margined with reddish-buff and breast somewhat spotted.

Common throughout Britain, except in Cornwall and some parts of Scotland. In Ireland breeds sparingly in almost every northern county, although (with exception of North Kilkenny) not recorded from the south, except on migration. Often arrives on our southern coasts about mid-April, but does not reach the north before May; even near London I have never seen it before April 22nd. On the South Downs, nest (neatly and compactly composed of moss and dry grasses, and lined with finer dry grasses) is placed in a grass tuft on a hillside, or among furze-bushes; laying commences as early as May 7th, a second brood being hatched in July. **Eggs**: 5 or 6, occasionally 7; bluish-green, usually slightly speckled at larger end with reddish; size ·75 by ·60. Call-note: a sharp *u-tick-tick*, accompanied by a slight movement of the tail; usually uttered while perched, but also on the wing. The not unpleasant song (delivered from a post or hedge) is a short warble uttered twice, each time preceded by a somewhat harsh note.

15. Pratincola rubicola (Linn.). STONECHAT.

Hab. W. and S. Europe, north to Scotland; eastward not found to north of Germany. Partially migratory in winter.

Male: head, nape and throat, black; sides of neck pure white; feathers of mantle black, edged with brown; wings and tail-feathers dark brown, with a very noticeable white patch on coverts of wings; upper tail-coverts white, mottled with dark brown, breast of a bright ruddy hue, paler on abdomen; under tail-coverts mottled with black and white; bill and feet black. Length 5·25; wing shorter and rounder than Whinchat's, and with 4th primary longest, instead of 3rd. In winter paler below, and upper feathers more margined with brown. Female: above brown, with darker streaks; rump reddish-brown; white patch on wing less conspicuous; throat mottled with black; breast much duller. Young resemble female, but have no black on throat.

Common generally, north to the Hebrides, although local in some parts. In Ireland a fairly common resident.

Migrates from the north of our island in winter, although in most parts of the South a proportion remain through the year. On South Downs nest-building commences early in April, but in the north a little later. Nest: among furze or on a bank among coarse herbage, sometimes among long grass in meadows; not very substantial; composed of moss and dry grass, lined with hair, feathers, vegetable down or wool; lining distinguishes it from Whinchat's. Eggs: 5 or 6; bluish-green (more *green* than Whinchat's and much paler) with distinct specks of pale reddish-brown; size ·72 by ·60. A second brood is produced about end of June, at which time the simple but pleasing song of male ceases. Most usual note in the breeding season is a sharp *whit-chack*, uttered from a furze-spray or on the wing. Food: grubs, worms, beetles, and winged insects, latter being commonly taken on the wing after the manner of a Fly-catcher.

GENUS V. RUTICILLA, *C. L. Brehm (1828)*.

Bill moderate, slender, gape with developed bristles; wings moderate; 3rd, 4th, or 5th quill longest. Tail moderately long. Tarsus slender, moderately long.

16. Ruticilla phœnicurus (Linn.). REDSTART.

Hab. Europe (except extreme south) and Asia, as far east as the Yenesei. In winter southward to Africa and Persia.

Male: upper plumage slate-grey; forehead and stripe above eye pure white; wing feathers brown with pale margins; upper tail-coverts and outer tail-feathers chestnut red; two middle tail-feathers brown; throat and sides of head deep black; breast and under wing-coverts chestnut; flanks brown; bill black; tarsi dark brown. Length 5·30; wing with 3rd primary longest. Female: chiefly greyish-brown;

chestnut-red of tail less conspicuous and black of throat and white forehead absent. Young birds in first year's plumage resemble female, but on leaving the nest are spotted like newly-fledged Redbreasts.

A regular summer visitor, arriving about end of April, and leaving in September. Breeds sparingly as far north as Sutherland ; rare in some parts of Wales and Cornwall ; in Ireland very rare, but has bred in Wicklow and Tyrone. Nest : in holes and hollows of trees, at a moderate height ; sometimes in holes of walls ; carelessly composed of moss, roots and dry grasses, lined with hair and feathers. Eggs : frequently 6, but I have seen 8 ; pale blue, very noticeably paler than Hedge sparrow's ; rarely speckled ; size ·73 by ·55.

17. Ruticilla titys (Scop.). BLACK REDSTART.

Hab. Central and Southern Europe, east to southern Russia ; also Asia Minor, Palestine, and (locally) North Africa. In winter south to Africa.

Male : upper plumage deep slate-grey ; wings dark-brown with a very noticeable white patch upon secondaries ; upper tail-coverts and tail (except two central feathers) chestnut-red ; forehead, cheeks, throat and breast black ; abdomen slate-grey ; under tail-coverts bright buff ; bill and feet black. Length 5·75 ; wing with 4th primary longest. After summer moult, black feathers of lower parts are somewhat margined with grey. Adult female : resembles female of common redstart but is much greyer both above and below, and under wing-coverts are grey in place of chestnut-buff. Immature birds are like female but young males have an indistinct whitish patch on wing.

According to the B. O. U. list, has bred in Notts ; has also probably nested in Essex (Zool., 1888, p., 390) and other counties. A regular winter visitor, in small numbers, to our eastern and southern coasts ; occurs sparingly also in south and east of Ireland ; to Scotland, a rare visitor.

GENUS VI. CYANECULA, *C. L. Brehm (1828).*

Differs to no very appreciable extent from *Ruticilla* although the individuals of this genus approach in appearance to *Erithacus*, the nidification also proving a close affinity to latter genus.

18. Cyanecula suecica (Linn.). RED-SPOTTED BLUETHROAT

Hab. Northern Palæarctic region, breeding from Scandinavia to Kamschatka. In winter migrating to Africa, India, and China.

Male : upper plumage warm dark-brown, with a white streak above eye ; upper tail-coverts and basal half of tail (except two central feathers) chestnut-red, terminal half dark-brown ; throat and upper breast bright blue, with a large central patch of chestnut-red ; lower breast zoned with black, white and chestnut ; abdomen buffish-white ; under wing-coverts bright buff ; bill black ; tarsi brown. Length 5·80. Female : under parts buffish-white, with a dark brown breast, and usually with one or two specks of blue. Immature birds resemble female.

Of irregular occurrence in winter on east coast, especially in Norfolk where small parties or even flocks occur almost annually, usually in September. Five or six stragglers have been taken in Scotland, but it is not recorded from Ireland.

19. Cyanecula wolfi (Brehm). WHITE-SPOTTED BLUETHROAT.

Hab. N. W. Africa and Western Europe, breeding northward to France, the Netherlands and North Germany. In winter migrating southward.

This form is considered by some to be of doubtful specific distinction from the last. Mr. Saunders (who goes so far as to include the two forms under one heading) denies that the present race had been proved to occur in this country prior to date on which he wrote (1888); he questions validity of Hancock's specimen said to have been taken near London in 1845 (Birds of Northumberland, p. 67) states that the bird with a white spot observed in the Isle of Wight from 1865 to 1867 (Harting, Handbook p. 104) was entirely blue-throated, and makes no reference to the Scarborough example (Zool., 1876, p. 4956).

It is distinguished from *C. suecica* by having patch in centre of blue throat white instead of red, but there is also a third and very rare form having entire throat blue. This last (which does not seem to have occurred with us) is not, however, usually considered separable from the white-spotted race.

GENUS VII. ERITHACUS, *Cuvier (1800)*.

Bill moderate, narrow, base depressed. Wings moderate, rounded, 4th, 5th, and 6th quills longest and nearly equal. Plumage lax.

20. Erithacus rubecula (Linn.). REDBREAST.

Hab. Europe, also N.W. Africa, Madeira, and the Azores. In winter partially migratory. Male: plumage soft; above olive brown; sides of neck bluish-grey; forehead, throat and upper breast, orange-red; centre of abdomen white; flanks and under tail coverts brownish; bill black, tarsi brown. Length about 5·80. Female does not differ appreciably from male. Newly-fledged young have feathers spotted with dull yellow, and with blackish brown tips; throat and breast being lighter than upper parts; belly dull white; bill brown; tarsi yellowish.

One of the most familiar of British birds. A resident species, breeding commonly even as far north as the

Orkneys and Hebrides, also in all parts of Ireland. Nesting commences early, first eggs being laid during latter part of March. Nest, composed of dead leaves, moss and a little grass, lined with hair and, perhaps, a few feathers, is placed in low banks, ivy-covered walls or holes in trees, in the latter case even at a height of 8 or 9 feet from ground. Eggs: 5 to 7; white, spotted about larger end with pale red; markings sometimes taking form of distinct blotches or being confined to a narrow zone around larger end; unmarked eggs are not infrequent; size ·80 by ·60. Two or three broods are produced. The sweet and musical song is heard during greater part of year.

GENUS VIII. DAULIAS, *F. Boie (1831)*.

Wings moderate, third quill longest. Tail rounded. Tarsus rather long and slender, having in front one long plate and four smaller scutellæ, instead of three as in *Erithacus*.

21. Daulias luscinia (Linn.). NIGHTINGALE.

Hab. Western and Southern Europe, also North Africa and Asia Minor. In winter migrating southward.

Male: upper plumage russet-brown; upper tail-coverts and tail dull chestnut; breast and flanks brownish-buff; belly dull white; bill and tarsi brown. Length 6·50. Female identical. Newly-fledged young have feathers of upper parts finely streaked with buff, and lower parts mottled with greyish-brown.

Arrives about middle of April, leaving in September. Common and generally distributed throughout south-east of England; an annual visitor to valley of the Trent, but very rare in Yorkshire, and unknown farther north; in south-west reaches mid-Devon; in Wales has only been found breeding in Glamorgan and Brecon. Chief haunts are woods, copses and tangled hedgerows in timbered meadows. Nest built during May among bushes and rank herbage, either *on* the ground or near to it in evergreens or other bushes; very loosely constructed, firstly of a layer of oak leaves, within which is a slight cup formed of horsehair, with a few grass blades and small roots. Eggs: 4 to 6; bluish-green, clouded uniformly over entire surface with warm brown, which produces an olive appearance; sometimes clouding is not uniform and so allows ground colour to become visible; size ·85 by ·60. The unrivalled song is heard (often at night) until the first or second week in June, when brood is hatched. Young are fed at first largely upon small caterpillars, later they frequent gardens.

Sub-family, Sylviinæ.

Young at first have in most genera pale or rufous margins to feathers of upper parts, but otherwise much resemble adults.

Genus IX. SYLVIA, *Scopoli (1769).*

Bill short, moderately stout, upper mandible slightly decurved. Wings moderate; first or bastard quill small, third or fourth longest. Tail fairly long, slightly rounded.

22. Sylvia cinerea, Bechstein. WHITETHROAT.

Hab. Europe, north in Scandinavia nearly to Arctic circle, east to Southern Russia and W. Asia. In winter south to Egypt and Abyssinia.

Male: head and nape dull grey; remaining upper plumage lightish brown; wings and tail darker; outer tail feather on each side mainly dirty white, next two also being tipped with white; secondary wing feathers widely margined with rust-colour; throat and centre of abdomen white; breast, flanks and under tail-coverts with a buffish tinge; bill brown above, flesh-coloured at base below; tarsi pale brown, claws darker. Length 5·60; wing 2·70. Female: duller in plumage and with the head and nape greyish brown. Young birds have a more decided reddish tinge above.

Very abundant from mid-April to September, except in extreme north of Scotland; common in every county of Ireland. Frequents hedges, copses and outskirts of woods; nettle-beds are favourite resorts, hence local name of "nettle-creeper." Nest: in small bushes on hedge-bottoms among nettles, etc., usually less than two feet from ground; very slightly constructed of dry grass-stalks, neatly lined with horsehair. Eggs: 4 or 5, seldom 6; yellowish or greenish-white, more or less finely spotted with light brown and with underlying purplish spots; sometimes markings consist of a few blotches around larger end; size, ·72 by ·55. Short song of male is not unpleasant, and is heard until the early summer; during pairing time it is frequently uttered on the wing, to the accompaniment of singular antics, as the bird shoots upward from its perch near the top of a small tree. Alarm-note, low and harsh.

23. Sylvia curruca (Linn.). LESSER WHITETHROAT.

Hab. Temperate Europe, north nearly to Arctic circle. In winter south to Persia, Arabia, and Northern Africa.

Male: crown dull grey; ear-coverts dark brown; upper plumage brownish-grey; wing-feathers browner, but quite without rust-coloured margins; tail brown, outer tail-feather on each side mostly white; throat and abdomen silvery white, with a faint pinkish tinge; flanks and under tail coverts tinged with buff; bill blackish; feet slate brown. Length 5·10. Female: browner above; length 5·00. Young scarcely differ from female.

Common throughout eastern and southern England, breeds sparingly as far north as southern Scotland, but farther north only occurs as a straggler, as is also the case in Devon, Cornwall, and most parts of Wales; in Ireland has once occurred, at Tearaght Rock Lighthouse, Co. Kerry, in October, 1890. Nest: shallower than that of *S. cinerea*; constructed of dried grasses bound together with spiders' cocoons and a little wool, lined with fine vegetable fibres and hair; in hedges or brambles at a height of from 3 to 5 feet from ground. Eggs 5 or 6; opaque, creamy white, sparingly spotted and blotched with brown and with underlying lilac spots. Size ·66 by ·50. Two broods appear to be reared. The insignificant song of male is heard both in spring and summer. The bird arrives about mid-April, leaving before October.

24. Sylvia orphea, Temm. ORPHEAN WARBLER.

Hab. South Europe, north to Southern France, east to South Russia; also Asia Minor and North Africa. In winter migrating to Africa.

Male: crown and sides of head black; upper plumage slate brown; secondary wing-feathers with pale margins; tail dusky-brown, but outer feather on each side is mostly white, and two next have whitish tips; under parts white, tinged on breast and flanks with brown; bill blackish; tarsi dark brown. Length about 6·00; wing about 3·00, Female: duller in plumage.

Two supposed occurrences are on record, *i.e.*, a female said to have been shot in 1848, near Wetherby, Yorkshire, and a young bird recorded by Mr. J. E. Harting as having been caught in Middlesex, in June, 1866.

25. Sylvia atricapilla (Linn.). BLACKCAP.

Hab. Whole of Europe, excepting extreme north; also Palestine and North Africa. In winter migrating to Africa.

Male: crown black; nape ash-grey; upper plumage greyish-brown; throat, breast, and flanks greyish; centre of abdomen white; bill brown; iris brown; tarsi slate. Length 5·70; wing with 3rd quill longest. Female: plumage browner than in male; crown reddish-brown. Young at first resemble female.

Arrives during third week of April, departing in September. Commonly distributed throughout England and Wales, becoming very scarce, however, towards north of Scotland; in Ireland breeds locally in nearly every county. Frequents woods, thickets, orchards, gardens, etc.; rather slight but compact nest, being placed in a small bush, clump of brambles, or tangled hedge, at height of from 1 to 4 or 5 feet; composed of dry grass, with some dead bracken, moss and spiders' cocoons, lined neatly with fibrous roots, dry grass, and horse-hair. Eggs: 4 or 5; some yellowish or greyish-white suffused with pale brown, and with darker nuclei or spots of the same; others almost uniform warm reddish brown, with a few darker mottlings and one or two blackish spots. Size ·75 by ·58. First eggs laid in May, and a second set in June or July. Song closely approaches in excellence to the Nightingale's. A few individuals sometimes pass the winter in the south of Ireland and England.

26. Sylvia hortensis, Bech. GARDEN WARBLER.

Hab. Europe, north to lat. 65°, but somewhat locally distributed. In winter south to Cape Colony.

Male: entire upper plumage light brown tinged with olive; an indistinct pale streak above eye; wing-quills dark brown with narrow pale margins; under parts whitish, tinged (except in centre of abdomen) with pale brown; bill horn-brown; iris hazel; tarsi slate-brown. Length nearly 6·00 Female similar. Young have broad pale margins to both secondaries and primaries.

Arrives about beginning of May, leaving usually before October. Of much more local distribution than the Blackcap, although covering much the same ground; very scarce in north of Scotland, also Cornwall and west of Wales; in Ireland rare and local, but has bred in at least six counties. Frequents similar places to the Blackcap. Nest: built in similar positions, but usually rather more bulky, and with more grass and less hair in its composition. Eggs: 4 or 5; white or greenish-white, mottled, clouded, or spotted (chiefly about the larger end) with several shades of brown, but quite without a red tint; some are blotched rather nicely with bright brown; size ·78 by ·60. Alarm note resembles Whitethroat's, but is less loud and harsh. Song mellow and pleasing, but of less compass than Blackcap's.

27. Sylvia nisoria (Bech.). Barred Warbler.

Hab. Central and south-eastern Europe; north to southern Scandinavia, west to the Rhine; also Persia and Turkestan. Migrating southward in winter.

Male: above ash-grey, turning to brownish-grey on wings; tail and wing-coverts barred with white; all except two central tail-feathers tipped with white and with white margins to inner webs; lower parts greyish-white, barred transversely with deeper grey; under wing-coverts mottled with white and grey; bill and tarsi brownish Length 6·25; wing 3.30. Female: browner; less barred. Young are said to show very few markings, except on the rump.

Seven occurrences (in autumn) are recorded, *i.e.*: one at Cambridge, identified by Prof. Newton, in 1879; four in 1884 (Yorkshire, Norfolk, Isle of Skye, and Co. Mayo), a second in Norfolk in 1888, and a second in Yorks in 1892.

GENUS X. MELIZOPHILUS, *Leach (1816)*.

Differs not greatly from *Sylvia*, in which it is merged by some authorities. Tail is longer and bill shorter and straighter; wings rather short, 4th quill longest. Feathers of crown are capable of partial erection.

28. Melizophilus undatus (Boddaert). DARTFORD WARBLER.

Hab. Western Europe (Spain, Portugal, France, and southern half of England), also Northern Africa.

Male: above dusky-grey; wings dusky-brown; secondary quills with pale margins; tail long, rounded, very dark brown, outer feathers margined and tipped with greyish-white; throat, breast, and flanks, reddish-chestnut; centre of belly white; bill orange at base, blackish at tip; tarsi orange-brown; iris orange-red. Length 5·00. Female: browner above, and chestnut below is paler and chiefly restricted to throat, while in young it is almost absent.

Resident, but scarce north of Thames; rather common in furze districts of Kent, Surrey, Sussex and Hampshire; slightly less so in the north-west counties as far as Cornwall. Has been reported as breeding in some eastern and midland counties, while I believe that it did so near Middlesborough, Yorkshire, in June, 1895 ("Ornithologist," April, 1896); Mr. C. Dixon has also recorded it as nesting in the south of the same county. Nest (similar to Whitethroat's, but more substantial) placed low down in dense furze; composed of slender twigs of furze with grass blades, moss and a little wool, lined with finer grass. Eggs: 4 or 5; greenish white, spotted closely with dark olive-brown and with a few underlying lilac spots; much like eggs of common Whitethroat, but markings are more distinct; size ·68 by ·50. First eggs are laid at end of April, there being a second brood in June or July. Food: chiefly insects, with some berries. Flight is undulating, but very short. If watched, skulks most successfully among the thickest furze.

GENUS XI. REGULUS, *Cuvier (1800).*

Bill slender, moderate, straight. Wings fairly long; first quill well developed, fifth longest; tail slightly forked, the feathers pointed. Tarsus fairly long, slender; claws hooked.

29. Regulus cristatus, Koch. GOLDCREST.

Hab. Whole of Europe, north to Arctic Circle; also temperate Asia eastward to the Amoor. Partially migratory in winter.

Male: forehead and cheeks greyish white; crest yellow in front, rich orange behind, bordered in front with dark brown and on each side with black; nape and back olive green; wing quills greyish brown with yellowish margins; secondaries black on lower parts with whitish tips; greater and lesser wing coverts tipped with white, forming two bars; tail feathers greyish brown with yellowish margins; under parts buffish white; bill brownish black; iris hazel; tarsi brown. Length 3·60. Female: duller in plumage; crest pale yellow. In young birds crest is lacking.

A fairly common resident, breeding throughout the British Isles, although local in some districts. In spring frequents woods and plantations where coniferous trees and evergreens are to be found, but in winter is more generally distributed. Nest: usually suspended *below* end of the horizontal branch of a fir, yew, or similar tree; height up to about 10 feet; constructed of green moss, bound together with fine grass, spiders' webs, wool, etc.; lined with a quantity of small feathers. Eggs: 5 to 10, laid in April or May; yellowish or buffish white, speckled round larger end with reddish-brown; size ·54 by ·40. Feeds upon insects which it seeks in low bushes, as well as upon gnarled trees; in winter moves with small bands of titmice, etc. Call note: a shrill weak *tzit, tzit.* Low sweet song of male is a repetition of *chivvit, chivvit, chivvit, chivvit.*

30. Regulus ignicapillus (Temm.). FIRE-CREST.

Hab. Southern and Western Europe, north and east to Germany and Southern Russia, also Asia Minor and Northern Africa. In winter partially migratory.

Male: forehead yellowish; crest rich orange, bordered on each side with black, below which is a white line, and below this again a narrow black streak passing through the eye and bordered beneath by a second white line, separating it from an indistinct blackish line which starts from the gape; nape yellowish-green; mantle olive-green; wing-feathers brown, margined with yellowish-green; greater coverts with a bar of white across tips; tail-feathers brown, with yellowish margins; lower plumage brownish-white; bill black; tarsi brown. Length 3·75. Female has crest pale yellow. Young lack crest, but show a distinguishing blackish line through eye.

A casual winter visitor; occurs almost annually on the south coast of England; unknown in Ireland and almost so in Scotland.

GENUS XII. PHYLLOSCOPUS, *Boie (1826).*

Bill slender, short, upper mandible a little discurved and indistinctly notched at tip. Wings moderately long, 1st quill developed, 3rd or 4th longest. Tail a little forked. Tarsus fairly long, claws curved.

31. Phylloscopus superciliosus (Gmelin.). YELLOW-BROWED WARBLER.

Hab. N.E. Siberia, from Pacific coast west to the Yenesei, north to within Arctic Circle. In winter south to N.E. India. A rare straggler to Northern Europe.

In autumn upper parts are yellowish green with a pale streak down centre of crown; on side of head a blackish line passes from base of bill through eye with a yellowish stripe above it, and a short streak of same colour beneath; under plumage pale yellow; wing-feathers dark brown, margined with pale yellow, and with broad tips of same to both greater and lesser coverts; bill and tarsi brown. Length about 4·00.

Five occurrences (in autumn) are recorded, viz., one shot by the late John Hancock in Northumberland (1838); a second recorded by Gould as taken near Cheltenham (1867); a third taken in Shetland Islands (1886), and identified by Mr. Harve-Brown; a fourth obtained in Co. Kerry, Ireland (1890), and now in Mr. R. M. Barrington's possession; and a fifth shot by Mr. G. H. Caton Haigh in Lincolnshire (1892).

32. Phylloscopus rufus (Bech.). CHIFFCHAFF.

Hab. Europe, north to Arctic circle, east to Central Russia. In winter south to Northern Africa and Persia.

Male: upper plumage olive-green, with a slight yellowish tinge, most pronounced on rump; from base of bill to above eye, a dull yellowish line; wing and tail-feathers slate-brown, margined with olive-green; whole lower plumage dull yellowish-white; under wing-coverts pale yellow; bill brown; iris hazel; feet blackish-brown. Length 4·70; wing 2·40. Female identical. Young lack yellow tinge.

Arrives about end of March, leaving again before October; exceptionally remaining through winter in south of England and Ireland. Generally distributed throughout England and southern Scotland, but local in some parts, as in Norfolk and other of our eastern counties; very rare in north of Scotland; breeds in every county of Ireland. Frequents chiefly woods and copses, but also abundant on commons, if unexposed, and in timbered meadow land. Usually sings in the tree-tops and seems chiefly to seek there for the insects and small larvæ upon which it subsists. Nest: placed on or near the ground, among brambles, coarse herbage, etc.; commonly on tangled banks. It is dome-shaped, with a wide entrance at side, and constructed of moss, dead leaves, and dry grasses, lined internally with a little hair and a quantity of feathers.

Eggs: laid in May; 5 or 6; white or creamy-white, but quite opaque, spotted with reddish-brown, usually in small very dark specks around larger end, but sometimes more distributed in larger aud paler spots; size ·60 by ·47. Tremulous, though pleasing little song may be syllabled as *chiff-chaff, chiff-chaff, chiv-chave;* it is heard continually in early spring, and is frequently resumed in August.

33. Phylloscopus trochilus (Linn.).

WILLOW-WARBLER.

Hab. Europe (excepting south-west portion from Balkan States to South Russia), north to North Cape. Also Siberia, west of the Yenesei. South in winter to the Mediterranean and Africa.

Male: upper plumage dull olive-green, with a distinct yellowish tinge, especially on rump, and with a pale yellow streak over the eye and ear-coverts; wing-feathers olive-brown, with yellowish-green margins; tail dull brown; whole lower plumage very pale sulphur yellow; under wing-coverts yellow; bill brown; iris hazel; tarsi pale brown. Length almost 5·00; wing 2·60. Female identical. In autumn yellow tinge is more pronounced.

Very common, arriving early in April and leaving late in September, known exceptionally to remain through winter in some districts. Frequents similar places to Chiffchaff, but is less restricted to woodlands; nest also placed in similar situations, but rarely on ground, and often 18 or 20 inches from it in low bushes, etc.; of similar shape, but usually constructed only of dry grasses externally, and lined with a quantity of feathers.

First eggs, 5 to 8, laid in May; white, marked in similar fashion to Chiffchaff's, but with much paler red, while shell is slightly transparent and never of an opaque white like that of Chiffchaff's eggs; size ·63 by ·47. A second brood is hatched. The simple little song is sweet and lively; the bird usually frequents bushes or the lower branches of trees; feeds chiefly upon aphides and other insects.

34. Phylloscopus sibilatrix (Bech.).

Wood-Warbler.

Hab. Temperate Europe; north to Scotland, Germany, and South of Sweden, south-east to Turkey, South Russia, and the Caspian; unknown in the south-west. In winter south to Africa.

Male: upper plumage yellowish-green, with a broad brimstone-yellow streak above and behind eye; wing and tail-quills slate with yellowish edges; throat and breast pale yellow; belly and under tail-coverts white; bill and tarsi brown. Length 5·0; wing 3·00 (wing is longer, tail shorter, and plumage yellower, than in the two preceding species). Female identical; young somewhat yellower.

A summer visitor to most counties of England and Scotland, but very local. Rare in Ireland; Mr. Ussher says it is a regular summer visitor to Powerscourt, Co. Wicklow, and Clonbrock, Co. Galway. Arrives late in April, departing in September; frequents chiefly the older woodlands and forests. Nest (placed on ground among rank herbage or in a bank) is similar to Willow-Wren's but lined with fine grass and hair—never with feathers. Eggs: 5 to 7; white, speckled all over with dark purplish-brown and lilac-grey, most closely at larger end; size ·65 by ·56. Call note, *tee-ur*; the "shivering" song is not easily syllabled.

GENUS XIII. AEDON, *Boie (1826).*

Bill rather long, strong, ridge of upper mandible curved and tip compressed. Wings with 1st quill small, 3rd longest. Tail long, rounded. Tarsus rather long; toes small.

35. Aedon galactodes (Temm.). RUFOUS WARBLER.

Hab. Northern Africa, Southern Palestine, and southern half of Spanish Peninsula. Migrating southward in winter.

Male: above rufous-brown, with a whitish streak above and behind eye; wings brown, with pale margins; tail reddish-chestnut, with a black band across end, succeeded by white tips to all but two central feathers; lower plumage white, tinged with buff on breast and flanks; bill and tarsi brown. Length 6·70.

Three occurrences only of this species have been recorded, viz., one shot near Brighton (1854), and two in Devonshire (1859 and 1876).

GENUS XIV. HYPOLAIS, *Brehm (1828).*

Bill moderate, tolerably stout, widened at base. Wings fairly long, 1st quill very small, 3rd longest. Tail moderate. Feet rather small.

36. Hypolais hypolais (Linn.). ICTERINE WARBLER.

Hab. Northern and Central Europe, south to Italy, north to Arctic Circle in Norway, west to the Rhine and Netherlands, east to Ural Mountains. South in winter to Africa.

Male: above olive-grey, with a yellow streak above eye; tail-feathers and primaries brown; secondaries with wide margins and tips of dirty white; lower plumage pale yellow; bill brown, base of lower mandible yellow; tarsi slate. Length 5·10; wing 3·00. Female scarcely differs.

A very rare straggler to our shores; five examples have been obtained, viz.: one near Dover (1848), one in Co. Dublin (1856), two in Norfolk (1884 and 1893), one near Newcastle-on-Tyne (1889). Also recorded as observed in Wicklow and Pembrokeshire in 1886.

GENUS XV. ACROCEPHALUS, *Naumann (1819).*

Bill moderately long, nearly straight, ridge of upper mandible arched, base widened, tip compressed. Wings somewhat short; 1st quill very small, 3rd longest. Tail rather long, rounded. Feet and claws large and strong; tarsus long.

37. Acrocephalus streperus (Vieill). REED-WARBLER.

Hab. Europe, north to south of Sweden; also Persia and Turkestan. In winter south to Africa.

Male: entire upper plumage pale brown, with a chestnut tinge which is most decided on rump; above eye a buffish-yellow streak; breast, flanks and under tail-coverts buffish-white; throat and belly white; bill horn brown above, base of lower mandible yellow; iris brownish-yellow, tarsi dark slate-brown. Length 5·50; wing 2·50. In autumn: more rufous above and more buffish below. Female scarcely differs, but is slightly smaller.

Arrives towards end of April, leaving in September. Common throughout England, except in the most northern counties, Cornwall and the west of Wales; of very casual occurrence in Scotland; in Ireland one is said to have been taken near Dublin, December 21st, 1843 (Thompson). Frequents chiefly reed-lined rivers, lakes and large ponds; nest being generally suspended between three or four reeds which pass through its sides; sometimes placed in willow, or other trees; it is very deep and neatly constructed of grass-blades, fine roots, moss and wool, lined with fine grass, hair, wool or feathers. Eggs 4 or 5; greenish white, variably spotted and suffused with dark olive-green and light brown, and with one or two black spots; size ·73 by ·53. Song quick and varied, often heard after dusk.

38. Acrocephalus palustris (Bech.).

MARSH-WARBLER.

Hab. Temperate Europe (excepting, apparently, west of France and Spanish Peninsula), north to Denmark, east to Turkestan. In winter south to Africa.

Adult: upper plumage *olive brown* with a nearly imperceptible eye streak; wing-quills with pale buffish margins; breast, flanks and under tail-coverts pale yellowish buff; throat and belly whitish; bill dark brown above, pale below; best authorities say that tarsi and feet are "brownish flesh-colour" during life. Length 5.50; wing 2·90. Upper parts are "earthy brown in summer, with a scarcely perceptible shade of rufous after the autumn moult, slightly paler on the rump" (Seebohm).

A spring visitor to southern counties of England, but its distribution is as yet undetermined owing to long confusion with the last species. It nests annually near Taunton in Somersetshire and a nest has been found recently near Bath (Zool., 1894, p. 304); has also been said to have bred on one, or perhaps two, occasions near Gloucester, while Mr. Saunders has examined a nest and eggs taken some years ago in Cambridgeshire. A nest and eggs taken near Banbury, Oxfordshire, is in Natural History Museum. The bird has also been identified in other districts. Nest appears never to be suspended over water, and seems to be usually found among osiers or in rank herbage near the water. Eggs: 4 to 7; clouded and spotted with olive-brown and with underlying spots of grey; ground tint whiter and markings less numerous than in Reed Warblers' eggs; size ·73 by ·55. Song is described as more attractive and of much greater compass than Reed Warbler's.

39. Acrocephalus turdoides (Meyer).

GREAT REED-WARBLER.

Hab. Europe, south of Baltic and English Channel; also Asia Minor and North Africa; migrating southward in winter from Europe.

Male: upper plumage olive-brown with an indistinct whitish streak over eye; wing and tail feathers with buffish margins; tail graduated or wedge-shaped; under plumage pale brown, whitish on throat and centre of belly; bill yellowish-brown, darker at tip; tarsi brown. Length 7·80; wing 3·80. Female similar, but a little smaller.

A very rare visitor. Mr. Harting (p. 14) says:—"The nest has been found in Surrey, Kent, Hants, Herts, and Northamptonshire." No other authority, however, appears to corroborate this statement. Mr. Saunders (p. 75) remarks that none of the reports of its breeding are authenticated; he reduces also the occurrences of the bird to eight, the first being the example obtained near Newcastle in 1847, and the rest being allotted to Kent, Essex, Hampshire, and Norfolk.

40. Acrocephalus aquaticus (Gmel.).

AQUATIC WARBLER.

Hab. Central Europe, north to N. Germany and S. Denmark, west to Central France, east to Central Russia; also North Africa. In winter south to Africa.

Mature bird has a distinctive streak of pale buff along the centre of the crown, bordered on each side by a blackish streak, which is succeeded again by a buff stripe above the eye; ear-coverts pale brown; nape and back buffish-brown, streaked with black; rump more rufous; lower parts yellowish-buff; throat and flanks with fine streaks of brown; bill brown above, paler below; tarsi yellowish-brown. Length 4·75.

Three examples are known to have been taken, viz., one shot near Brighton (1853), a second taken near Loughborough (1864), and a third shot near Dover, prior to 1871.

41. Acrocephalus phragmitis (Bech.).

SEDGE-WARBLER.

Hab. Europe (excepting S.W. portion), north to lat. 70° in Norway, but scarce in the south, east to Ural Mountains, south to N. Italy; also Siberia, east to the Yenesei, and Palestine.

Male; crown pale brown, streaked with dark brown, and bordered by a noticeable yellowish-white streak above each eye; ear-coverts brown; nape and mantle reddish-brown, suffused with dark brown; rump and upper tail-coverts tawny-brown; wing-quills and tail dark-brown, with pale margins; throat white; remaining under parts pale buff, deeper on flanks; bill dark brown above, yellowish at lower base; iris brown; feet pale brown. Length 5·00. Female: less rufous, and duller in general tints. Young: slightly spotted with light brown on breast.

Arrives about third week in April, leaving at end of September; breeds in every county in Britain and in most parts of Ireland. Frequents rivers, lakes, ponds; also ditches and hedgerows in damp meadows, &c. Nest is usually placed in low bushes or brambles among coarse herbage; usually low down, but I have seen it at 8 feet from ground; exceptionally suspended like a Reed-Warbler's between three or four tall sedges, etc., but not over water; rather deeply constructed of moss, grass, and a little wool, lined with fine grass and horsehair. Eggs: 4 to 6; pale brownish-yellow, mottled and clouded with a darker tint and with a few superficial black streaks; size ·68 by ·52. The loud, persistent, hurried, and chattering song is heard at night as well as by day.

Genus XVI. LOCUSTELLA, *Kaup (1829).*

Differs not greatly from *Acrocephalus.*

42. Locustella nævia (Boddaert).

GRASSHOPPER WARBLER.

Hab. Europe, north to the Baltic and British Islands, occasionally to Norway: of local distribution. Migrating from North Europe in winter.

Male: upper plumage pale greenish-brown, each feather (except upper tail-coverts) having a dark brown centre; large, slightly-graduated tail, and the wing-quills brown with narrow greenish margins; under plumage pale yellow, tinged and speckled on breast with dark brown; under tail-coverts very long, yellowish-white, streaked with dark brown; bill dark brown above, yellow beneath; feet large, yellowish-brown. Length 5·30. Female: identical.

Distributed through England, Wales, the southern half of Scotland, and in many Irish counties, but very local. Nest: usually in a tussock of rank grass or among furze and rank herbage on commons, the outskirts of copses, or at the edge of a ditch; composed of a quantity of dry grass with a little moss, lined with fine grass. Eggs: 4 to 7; pale reddish-white, speckled all over with reddish-brown, most thickly around larger end. Size ·71 by ·54. First eggs laid about end of May; there is a second brood. Remarkable song is a clear, sustained trill, resembling, at a distance, the continued sound of a tiny bell; it is always uttered when the bird is perched motionless in a small bush and is usually continued for about a minute without a pause. Arrives at end of April, leaving in September.

43. Locustella luscinioides (Savi). Savi's Warbler.

Hab. Central and Southern Europe, north to Holland, west to Southern France and part of Spain, east to the Caspian; also Algeria. Migrating southward in winter.

Adult bird has upper plumage reddish-brown; throat white; breast, flanks and under tail-coverts, pale buffish brown; belly whitish; bill brown, yellowish at lower base; tarsi pale brown; claws darker. Length 5·60; tail broad and rounded; Mr. Saunders says it "shows in certain lights some faint transverse bars."

Formerly bred in small numbers in East Anglia, but probably ceased to do so about 40 years ago, since which time there is no record of its occurrence.

Sub-Family Accentorinæ.

GENUS XVII. ACCENTOR, *Bechstein (1802).*

Bill strong, rather conical in shape, tip of upper mandible slightly notched. Wings somewhat rounded; 1st quill very small, 3rd usually longest. Feet strong; hind claw rather long.

44. Accentor modularis (Linn.). HEDGE-SPARROW.

Hab. Europe (except extreme south), north to lat. 70° in Norway Partially migratory in winter.

Male: head and nape bluish-grey, spotted with brown; feathers of back and wing-coverts reddish-brown, with blackish centres; greater coverts slightly tipped with white; primaries and tail dusky brown; throat and breast bluish-slate; belly whitish; sides tawny, with dark brown streaks; bill pale brown, darker at tip; iris reddish-brown; feet orange-brown. Length 5·50. Female: duller, slightly smaller, and more closely spotted on head and nape.

Resident throughout British Isles, north to Inner Hebrides. Familiar nest is placed near the ground in bushes and hedgerows; composed of green moss, small roots and dry grass, lined with hair and a little wool or feathers. Eggs, 4 or 5; immaculate blue; size ·78 by ·57. Two or three broods are produced, first eggs being laid at end of March. Song, short, but sweet and tender; it may be heard from February onward. Usual note, a low, plaintive chirp.

45. Accentor collaris (Scop.). ALPINE ACCENTOR.

Hab. Mountain ranges of Southern and Central Europe; also Asia Minor and North Persia; straggling northward in winter.

Adult: head and nape greyish-brown, streaked with darker brown; feathers of back brown, with dark centres; lesser and greater wing-coverts tipped with white, forming two bars; quills dark brown; secondaries tipped and edged with chestnut; tail feathers dark brown, with dull white tips; throat white, spotted with black; breast and centre of belly brownish-grey; sides mottled with reddish-brown; bill blackish above, yellow at lower base; tarsi orange. Length 7·00.

Fourteen examples have been recorded as taken or seen in England and Wales since 1817, when one was shot at Walthamstow, Essex, but not recorded until 1832.

Family Cinclidæ.

GENUS XVIII. CINCLUS, *Bechstein (1802).*

Bill of medium length, nearly straight, base laterally compressed. Gape narrow; bristles absent. Wings short, broad; tail short. Tarsus moderately long. Plumage close; whole body clothed with down.

46. Cinclus cinclus (Linn.).

BLACK-BELLIED DIPPER.

Hab. Scandinavia and Northern Russia, straggling southward and westward in winter.

Although accorded separate rank in the British Ornithologists' Union list, this form is considered by recent workers to be so doubtfully distinct from the next that I have included the latter as a sub-species only. The northern bird is distinguished by the upper belly being black instead of chestnut. The late Mr. Stevenson recorded this race as an occasional winter visitor to Norfolk and Suffolk, and Mr. J. H. Gurney has had an example from Yorkshire.

46a. C. cinclus aquaticus (Bech.). DIPPER.

Hab. Central and Western Europe, north to Scotland.

Male: head and nape deep brown; remaining upper parts blackish-brown, with dark greyish margins to the feathers; throat and breast white; upper part of belly dark chestnut, changing to black on the flanks and lower belly; bill and feet brownish black. Length nearly 7·00. Female: identical. Young: paler above, and no chestnut below.

Common throughout West and North of England, and whole of Scotland, north to the Hebrides; also most of mountainous parts of Ireland. Frequents running brooks and rivers, nest being placed in rocky banks or ledges under bridges, sometimes in boughs of trees over water. It is oval, with an entrance at side; composed of green mosses, lined with dry grasses and dead leaves. Eggs: 4 to 6; shell without gloss, pure white; size 1·00 by ·75. First eggs laid in March, there being two or three broods. Song, weak but sweet and pleasing; often heard in winter. Swims with ease, and has also a habit of seeking its food at bottom of pools and streams, walking right in from edge.

Family Panuridæ.

GENUS XIX. PANURUS, *Koch (1816).*

Bill short, somewhat conical, upper mandible a little decurved and longer than lower, which is straight. Wings with 1st quill very minute, 3rd longest. Tail long, the feathers much graduated. Tarsus rather long; feet stout; claws slightly curved.

47. Panurus biarmicus (Linn.). BEARDED TITMOUSE.

Hab. Southern and Western Europe, north in latter to Holland and England; also Turkestan and South Siberia. In winter partially migratory.

Male: crown blue-grey; upper parts and tail reddish-buff; wing-coverts banded with black, white and reddish-brown; primaries brownish, edged with white; a black moustache-like patch runs from before the eye to side of throat, where it terminates in a pointed tuft; throat dull white; breast with a slight tinge of flesh-colour; sides reddish-buff; under tail-coverts black; bill and iris yellow; feet black. Length 6·50. Female: duller than male; crown, moustache and under tail-coverts, pale orange-brown. Young: general reddish tint of male quite replaced by buffish-yellow; centre of back and outer tail-feathers chiefly black.

A resident, but only known now to breed in the Norfolk Broads, and (according to Mr. Saunders) in one locality in Devonshire. It formerly bred in many eastern and southern counties, and still occasionally straggles in winter to its old haunts. Nest: placed in aquatic herbage, or broken-down reeds; cup-shaped, composed of blades of sedges and grasses, lined with vegetable down. Eggs: 4 to 7; shell glossy, creamy-white, with short streaks of reddish-brown; size ·70 by ·56.

Family Paridæ.

GENUS XX. ACREDULA, *Koch (1816)*.

Bill very short, strong, compressed laterally, both upper and under mandibles curved, former longer than latter. Wings with 4th and 5th quills longest. Tail slightly longer than body, narrow and much graduated. Tarsus fairly long, claws rather curved.

48. Acredula caudata (Linn.). WHITE-HEADED LONG-TAILED TITMOUSE.

Hab. Scandinavia, Russia and Eastern Germany; also Siberia.

This is the true *Parus caudatus* of Linnæus, distinguished from the common British form by the head in adult birds being entirely white, while the general tints are somewhat brighter. Prof. Collett says that the *young*, in nestling plumage, have *dark eye-stripes*. It is a very rare wanderer to us. Mr. Harting has recorded the obtaining of two examples in the North of England.

48a. A. caudata vagans (Leach). BRITISH LONG-TAILED TITMOUSE.

Hab. Western Europe, north to Scotland, south to France, east to Belgium and the Rhine.

Male: forehead and crown white, margined on each side by a black stripe running from bill, over eye, to nape and upper part of mantle, which are also black; scapulars and back pale brownish-red; wing-coverts and quills black; secondaries edged with white; tail black, three outer feathers each side edged and tipped with white; under plumage white; abdomen and under tail-coverts with a rosy tinge; bill black; feet blackish-brown; claws black. Length 5·50. Female has white on crown more restricted.

Resident and fairly common. In winter wanders in small parties in company with other Titmice, Goldcrests, etc., feeding in the trees upon small insects and their larvæ. The beautiful nest, commenced late in March, is placed in holly or hawthorn bushes, sometimes high up in forks of oak or other trees, occasionally in furze; in shape oval, with a small entrance near top; composed of lichens, moss, and spiders' webs, lined with a quantity of small feathers. Eggs: 7 to 10; white, finely speckled with pale red-brown around larger end; size ·55 by ·43. Usual note a quick shrill *zee, zee, zee*. May be distinguished by its small size and long tail.

GENUS XXI. PARUS, *Linnæus (1766).*

Bill moderate, strong, straight and rather conical. Wings with 1st. quill short, 4th or 5th longest. Tail moderate, nearly even. Tarsus moderate; feet strong; claws curved.

49. Parus major, Linn. GREAT TITMOUSE.

Hab. Europe, north to Arctic Circle in Norway; also Siberia south of lat. 58° N., and N. Africa, Asia Minor and Persia.

Male: head black, excepting cheeks and ear-coverts which are white; nape yellowish green with a whitish spot in centre; mantle olive-green; rump and wing-coverts blue-grey; greater coverts with white tips; quills dark brown with greenish margins; tail-feathers dusky, outer pair with white margins; throat, and a broad stripe down centre of breast and belly, black; sides dull yellow; under tail-coverts whitish; bill black; feet slate. Length 5·50. Female: duller, with a less distinct belly-stripe.

Common everywhere. Nest: placed in holes in trees, or in many curious situations, such as gate-posts, holes in walls, nests of larger birds, inverted flower-pots, etc.; consists of moss and grass with a cup-shaped lining of hair, wool and feathers. Eggs: 5 to 8; white, with spots and blotches of pale red; size ·70 by ·54. Two broods are produced, first eggs being laid in April. The loud, clear song has been well-imitated in the birds' local name of *sit-ye-down*, the three notes being repeated several times in quick succession. Alarm note a loud rattling *chee, chee, chee.* In winter it roves through the woods in small parties. I have heard it tapping on old trees like a Woodpecker and it will sometimes bore in decayed wood for insects.

50. Parus ater, Linn. CONTINENTAL COAL TITMOUSE.

Hab. Europe (excepting British Islands) north to lat. 65° N. in Norway. Migrating from extreme north in winter, when it straggles to Britain.

This, the typical form, differs from that resident in British Islands, in having back slate-grey, difference being most marked in North-European specimens. Examples have been obtained in Norfolk.

51. P. ater britannicus (Sharpe and Dresser). BRITISH COAL TITMOUSE.

Hab. British Islands.

Male: head black; nape white; cheeks white; throat black; mantle olive-brown; rump with a buffish tinge; wings greyish-brown with pale margins; greater and lesser coverts tipped with white; tail greyish-brown; breast whitish; flanks pale fawn colour; bill and tarsi brownish black. Length 4·20. Female: similar. Young: nape and cheeks yellowish; black on crown and throat less pure.

Distributed throughout British Isles; frequenting woodlands, but nowhere very abundant. Nest: placed in holes in trees or walls or sometimes in banks; composed of moss wool, hair and feathers. Eggs: 5 to 9; white, sparingly spotted with pale red; size ·61 by ·46; laying commences in April. The musical little song bears some resemblance to the Great Tit's, but is much less loud and distinct. When feeding utters a slight cheeping; call note of male a sharp chirp. In winter associates with other Titmice. The bird is most partial to fir woods.

52. Parus palustris dresseri, Stejn. MARSH TITMOUSE.

Hab. British Islands.

Male: cap and nape black; cheeks dull white; chin and

throat black; upper plumage olive-brown, with a warmer tinge on rump; wing feathers and tail greyish-brown, with pale outer margins; under parts buffish-white; bill black; iris dark hazel; feet blackish. Length almost 4·50. Female, identical. Young, browner.

The description is that of the British race, *P. palustris dresseri* of Stejneger, which, however, is not usually separated by our ornithologists. The typical *P. palustris* of the Continent is olive-grey above; it is an intermediate form, the other extreme being *P. palustris borealis*, of Arctic Europe, which is larger, light grey above, and white below.

Resident and tolerably common in England and Wales; rare in Scotland, only breeding in the south; in Ireland has occurred in Antrim, Kildare, and Dublin (Thompson), but Dr. Trumbull tells me it has not been obtained for many years, and he doubts if it has ever bred. In winter it is very common in the woodlands around London in company with other Titmice, but in the breeding season quite rare. I have always known it to excavate for itself a nesting hole like the Lesser Spotted Woodpecker's, in a decayed willow or other stump, a few feet from the ground, but it is said to also choose a ready-made site. Nest: a little moss, wool, and hair. Eggs: 5 to 8, white, spotted about larger end with reddish-brown; size ·63 by ·48. Song is an almost inaudible *sis, sis, sis, see;* usual note when feeding a weak *zee, zee, zee;* call-note a shrill quick *sissle-chip*; when alarmed it utters a shrill squeaking, followed by a harsh *cree, ree, ree, ree.*

53. Parus cæruleus, Linn. BLUE TITMOUSE.

Hab. Europe, east to the Ural Mountains, north to 64° N. lat. in Norway.

Male: feathers of crown bright blue, capable of partial erection; from forehead to nape (above eye) a white line, succeeded by a bluish-black line passing through eye, below which cheeks are dull white; lower nape dark blue, this colour extending below cheeks and joining blue-black throat; back yellowish-green; tail and wings blue; greater coverts and secondaries tipped with white; under parts dull yellow; bill black; tarsi slate. Length 4·50. Female: duller in plumage.

Common everywhere. Frequents woods, gardens, orchards, etc. Nest: usually in holes in trees or old walls, but many curious sites have been chosen; composed of moss, wool, hair, and feathers. Eggs: 7 to 12; white, spotted with pale red; size ·60 by ·45. First eggs are laid late in April; the female is an obstinate sitter, and hisses if disturbed; young are fed chiefly with small caterpillars. Song is a shrill, quick *chirrif-ee*, *chirrif-ee*, *chirrif-ee*, commenced at end of January; when alarmed bird utters a harsh rattling note; associates with other species in winter.

54. Parus cristatus, Linn. CRESTED TITMOUSE.

Hab. Western and Northern Europe, north to lat. 64° in Norway, east to Central Russia, south to Spain; not found in S.E. Europe.

Male: feathers of head black, edged with white, lengthened and directed upwards, forming a noticeable crest; cheeks white, mottled with black and bordered behind by a crescentic streak of black, behind which is a collar of white, followed by another black streak running from nape and joining black of throat and upper breast; back and wing-coverts olive-brown; primaries and tail dusky-brown; under parts white, tinged with pale brown; bill black; tarsi blackish-slate. Length barely 4·70. Female: less black on throat and more dull white on crest, while latter is also smaller. Young have very little crest.

Practically confined to N.E. of Scotland, having bred only in Moray, Ross, Inverness, Elgin, Banff, Aberdeen, and (probably) Perthshire; a rare straggler to South Scotland. In England examples have been recorded from Cumberland, Durham, Yorkshire, Suffolk, Kent, Middlesex and Hants, and Mr. J. Whitaker tells me he has one taken in Notts. In Scotland breeding hole is bored in decayed fir-stumps; nest formed of moss, feathers, and rabbits' fur. Eggs: usually 5; white boldly spotted towards larger end with red; size ·65 by ·50: laid early in May.

Family Sittidæ.

Genus XXII. SITTA, *Linnæus (1766)*

Bill moderately long, straight, strong, somewhat conical. Tongue short, tip bristly; nostrils basal, concealed by small, hairy feathers; wings fairly long; tail short, even; tarsi short; feet strong; claws hooked.

55. Sitta cæsia, Wolf. NUTHATCH.

Hab. Temperate Europe, north to Baltic and England; also Asia Minor and North Africa.

Male: above blue-grey, with a black stripe from base of bill to side of neck, passing through eye; primaries and centre tail-feathers greyish-brown; outer tail-feathers blackish, tipped with grey and barred with white; chin and cheeks whitish; under parts buffish-yellow; flanks and under tail-coverts chiefly dark reddish-brown; bill blackish, paler at base; tarsi yellowish-brown. Length about 5·75. Female: plumage duller.

Resident and distributed over most parts of England and Wales, north to Yorkshire; rare in northern counties, a straggler to Scotland, and unknown in Ireland. Fairly common in older woods, parks, and forests, but very scarce away from these. Nest: commenced in April, usually in a hole in the trunk or larger limbs of a tree, the entrance being plastered with clay to reduce it to size of bird; nest merely a few leaves, flakes of bark, etc. Eggs: 5 to 8; white, spotted or blotched with reddish-brown; size ·78 by ·56. Food: beech-mast, acorns and nuts, latter being placed in a crevice and broken by repeated strokes from the bill. Frequents upper branches of tall trees, the male in spring uttering a loud whistling *tui, tui, tui.*

Family Troglodytidæ.

GENUS XXIII. TROGLODYTES, *Vieillot (1807).*

Bill moderately long, slender, pointed, very slightly decurved; wings short, rounded; tail short; tarsus rather long; hind toe large; claws long, hooked; plumage lax.

56. Troglodytes troglodytes (Linn.). WREN.

Hab. Europe, north to Sweden and North Russia; also Asia Minor, Persia, and extreme north of Africa.

Male: above light reddish-brown, with numerous transverse dark brown bars; above eye a dull whitish line; primaries barred with reddish-buff and blackish-brown on the outer webs; throat dull white; under parts pale brown, barred with darker brown on the sides; bill dark brown above, pale below; iris hazel; tarsi, brown. Length 3·75. Female: slightly smaller and much browner on throat and breast.

Resident and common throughout British Isles. Frequents exposed as well as wooded situations. Domed nest is placed in hedges, bushes, ivy on trees or walls, or in haystacks; in the last case I have found it to be made of hay externally, but it is usually built of moss and dead leaves; a lining of feathers is invariably added before the eggs are laid, but many nests are deserted before reaching that stage. Eggs: usually 5 to 7; shell, glossy, white, more or less speckled around larger end with pale red; size ·68 by ·50. The loud merry song may be heard nearly all the year. Usual note a quick *chit-chit-chit-chitr.*

56a. T. troglodytes hirtensis (Seebohm).
ST. KILDA WREN.

Hab. Island of St. Kilda.

This race is restricted to St. Kilda (off the west coast of Scotland), but since the late Henry Seebohm separated it specifically as *T. hirtensis* (Zool. 1884) it has probably been well-nigh exterminated. Few authorities regard it as a valid species, yet it is certainly entitled to sub-specific rank. It is much paler than the type, the general tint being greyish in place of reddish, causing the transverse bars on upper parts to be more pronounced; the bill is also slightly longer than in the type, and possibly more straight, while the bird itself appears to be appreciably larger.

Family Certhiidæ.

GENUS XXIV. CERTHIA, *Linnæus (1766).*

Bill tolerably long, slender, pointed, somewhat compressed and moderately decurved; wings moderate, rounded; tail pointed, of 12 long, stiff, tapering feathers; feet large; tarsus short; hind toe rather short; claws much curved; plumage lax.

57. Certhia familiaris, Linn. TREE-CREEPER.

Hab. Northern temperate regions of Old World, north to Arctic Circle, south to Mediterranean, Persia, the

Himalayas and Japan. Partially migrates from coldest regions in winter.

Adult: head, nape and mantle, rusty brown, each feather having a buffish white streak in the centre; rump and upper tail-coverts rust colour; wing-quills dark brown, edged on outer webs with reddish-buff, and (all but first three) crossed by a band of rusty-white, wing-coverts being also spotted with the same; tail feathers dark brown, edged with olive and tinged with rusty-brown, shafts being yellow; throat, breast and belly, silvery pearl-white; sides with a buff tinge; above the eye a short whitish streak; bill brownish-black above, yellowish-white below; feet pale brown. Length 4·25; wing 2·50; tarsus ·55. Young: much paler, general tint rusty buff; bill shorter and straighter.

Breeds throughout British Isles wherever old trees are found. It seems certain that numbers move southward from N. Britain during winter, as at this season it is very abundant, in company with Titmice, etc., in S.E. of England, while in spring but few pairs remain. Searches constantly upon tree-trunks for small insects and their larvæ, ascending from near the ground to the highest part, the stiff pointed tail being pressed into the bark as a support; flits from one tree to another with a quick, wavering flight. Upon old trees also it collects materials for its nest, which is placed in a crevice in a gnarled oak, behind loose bark on a decayed tree, or in a hole; composed of small dead twigs, chips of decayed wood and bark-strips, held together with spiders webs and lined with wool and smaller feathers of Titmice and other tree-haunters. Eggs: 6 to 8; white, boldly spotted around larger end with reddish-brown; size ·64 by ·47. Usual note is an almost inaudible *zee*, *zee*, but in spring male sometimes sings while ascending a tree, song being shrill and much like that of the Wren.

GENUS XXV. TICHODROMA, *Iliger (1811.)*

Bill slender, rather long, very slightly decurved. Tail moderate, square, not used when climbing. Wings broad, moderately long.

58. Tichodroma muraria (Linn.). WALL-CREEPER.

Hab. Southern-temperate regions of the Old World; inhabiting mountain ranges of Europe, north to the Vosges and the Carpathians; in Asia extending eastward to China.

An extremely rare straggler to England. One was shot, May 8th, 1872, in Lancashire and recorded by Mr. F. S. Mitchell. Previously Willughby (Orn. 1576) observes that it is said to have occurred in England, and one shot in Norfolk, in 1792, is found described in a letter from Robert Marsham to Gilbert White (Trans. Nort. and Nor. Nat. Soc., 1876). The adult is conspicuous from its having the wing-coverts and the basal halves of outer webs of primaries deep crimson.

Family Motacillidæ.

GENUS XXVI MOTACILLA, *Linnæus (1766)*

Bill moderate, slender, almost straight, tip of upper mandible indistinctly notched. Wings moderate; first primary, very minute, 3rd, or 4th longest; inner secondaries very long. Tail long, feathers almost equal in length. Tarsus rather long; claws moderate.

59. Motacilla alba, Linn. WHITE WAGTAIL.

Hab. Northern regions of Old World; in Europe north nearly to North Cape, south to Asia Minor and extreme North of Africa. Leaves North Europe in winter.

Male: forehead and sides of head and neck as far as

shoulders white ; crown and nape black ; mantle, wing-coverts and rump pale ash-grey ; lesser and greater wing-coverts tipped with white ; wing-quills brownish-black ; secondaries margined externally with white ; two outer tail feathers on each side chiefly white, remainder blackish ; chin, throat, and upper breast black ; lower breast and belly white ; sides tinged with grey ; bill and tarsi black. Length 7·00. Female : duller generally. After autumn moult both sexes have throat white but show a black band on breast. Immature birds have a slightly yellowish tint.

A scarce summer visitor, but has been recorded from nearly every county in England, while in Middlesex in particular it occurs regularly, and I believe breeds every year; has also been reported to have bred in most of the southern and eastern counties. A scarce visitor to Scotland. In Ireland two examples have been obtained by Mr. R. Warren (1851 and 1893) and two or three others have been observed. Nesting habits are similar to those of Pied Wagtail ; eggs can only be safely distinguished by identifying the birds.

60. Motacilla lugubris, Temm. PIED WAGTAIL.

Hab. N. W. Europe ; breeds in British Isles, N. W. France, and sparingly north-eastward to Belgium, Holland, and exceptionally S. Norway. Majority move southward in winter.

Male : forehead and sides of head pure white, this extending also down sides of neck, although more restricted here than in *M. alba* ; entire upper parts black ; both wing-coverts tipped and margined with white ; inner secondaries also margined with white on outer webs ; two outer tail-feathers on each side chiefly white ; chin, throat, and breast black ; centre of belly whitish ; sides black ; bill and tarsi black. Length 7·40. Female : dusky grey above with blackish streaks ; black on under parts more restricted; tail slightly shorter. In autumn, male becomes dusky grey on back, female rather paler, while in both sexes throat is greyish-white.

Common throughout British Isles, majority arriving in March and leaving at end of November, when I have seen parties of 30 or 40 assembled in Essex; a percentage remain through winter in the southern counties. Nest: built in April in banks, rough walls, quarries, roots of fallen trees, etc.; composed of small roots, grass, and moss, lined with hair, wool, or feathers. Eggs: 4 to 6; greyish or bluish-white, speckled all over with light brownish-grey; size ·80 by ·60. Flight-note is a shrill *chiz-ick, chiz-ick*, but in spring male utters a simple but pleasing song. I have often seen it take flies on the wing, but it is chiefly a ground-feeder.

61. Motacilla melanope, Pallas. GREY WAGTAIL.

Hab. Western and Southern Europe, north to Scotland, Germany, and Central Russia; also temperate Asia, eastward to Japan.

Male: upper parts slate-grey with a white line above the eye and a broad white stripe below, bordering the black chin and throat; wing-quills brownish-black; secondaries edged with dull white; upper tail-coverts yellowish, conspicuous in flight; outer tail feather on each side white, and also the greater part of the two next; remainder dark brown; under parts pale yellow; bill dark brown; tarsi light brown. Length about 7·30. Female: duller; tail shorter. Both sexes lose black throat in winter, and have a brownish tinge on the yellowish under parts.

Breeds sparingly in S.W. counties of England, throughout Wales, the north of England, and the whole of Scotland; also all over Ireland, although nowhere very numerous. Has bred occasionally in Sussex and Kent, but to east side of England generally is a winter visitor, being then seen flitting about the fields or paddling up and down in small streams and rivulets, uttering a shrill *chiz-ip*; soaring, or flight note, however, is a longer ***chip***, *chip*, *chip*, *chiz-ip*. Nest: placed in rocky banks of streams, etc.; composed of fine roots, moss, and grass, lined with much hair. Eggs: 4 or 5; greyish white, mottled with pale yellowish-brown; size ·75 by ·55.

62. Motacilla flava, Linn. BLUE-HEADED YELLOW WAGTAIL.

Hab. Central Europe, north to Holland and Britain. In winter migrating to Africa.

Male: crown and nape blue-grey; ear-coverts darker; above eye a white line; back rather dark yellowish-olive; wings dark brown, both greater and lesser coverts being tipped with yellowish-white and secondaries margined with the same; two outer tail-feathers on each side chiefly white; remainder dusky-brown; chin and lower cheeks white; throat, breast, and under parts, rich yellow; bill and tarsi black. Length 6.50. Female: duller above and below. Young: like those of *M. raii*, but are said to always show a *white* eye-stripe.

An irregular spring visitor to south and east of England. The late John Hancock recorded it as breeding on several occasions near Gateshead-on-Tyne, and it has without doubt done so in other localities. In Scotland has occurred two or three times, but has not been proved to visit Ireland. Nest and eggs scarcely differ from those of *M. raii*. Term "Grey-headed Wagtail," sometimes applied to this species, belongs properly to *M. viridis* of Gmelin, inhabiting Arctic Europe and also Siberia; this form has crown dark blue-grey, almost black, and eye-streak absent; has possibly occurred in England, at Penzance. It has, however, often been confused with *M. cinereicapilla*, of Savi, inhabiting the shores of the Mediterranean.

63. Motacilla raii (Bonaparte). YELLOW WAGTAIL.

Hab. Northern France and British Islands, also said to breed in the Caspian region. In winter migrating to Africa.

Male: crown and upper parts light greenish-olive; above eye and ear-coverts a pale yellow stripe; wings dark brown, coverts and secondaries margined and tipped with buffish-white; two outer tail-feathers chiefly white, rest dusky-brown; entire lower parts bright yellow; bill and tarsi black. Length 6·25. Female: above olive-brown; eye stripe and under parts paler yellow; breast tinged with brown. Young: like female, but even less yellow; sides of neck and breast spotted with brown.

Fairly common throughout Great Britain, excepting north of Scotland and the extreme S.W. of England; in Ireland breeds commonly around Lough Neagh, but seldom in other localities. It arrives early in April, usually leaving in September, but I have known it to occur near London in winter ("Birds of London,") and it has been recorded also from Bath on March 9th ("Ornithologist," April, 1896). Nest: placed on ground in meadows and cornfields, often concealed in grass by ditches and streams; composed of dry grasses and moss, lined with hair, feathers, etc. Eggs: 5 or 6; greyish white, mottled with pale yellowish-brown, and with one or two black streaks; size ·78 by ·57. First eggs laid early in May; two broods often reared. Flight-note is a shrill *chee, chee, chit-up*.

GENUS XXVII. ANTHUS, *Bechstein (1807).*

Differs very slightly from *Motacilla*. Tail is shorter and slightly forked. Tarsus also shorter; hind claw moderate and curved in some species, but in others long and nearly straight.

64. Anthus pratensis (Linn.). MEADOW-PIPIT.

Hab. Europe, north to Iceland, and lat. 70° in Norway, south to the Pryenees, Alps and Carpathians, east to beyond the Ural Mountains; also Asia Minor. In winter southward to Africa.

Male: above, rather light olive-brown with a narrow dark centre to each feather; primaries dark brown margined exteriorly with yellowish-green; wing-coverts and secondaries edged and tipped with buffish-white; outer tail-feather on each side mainly white, next one with some white near tip, rest dark brown; above eye an indistinct whitish stripe; throat dull white; lower parts buffish-white, streaked on sides of neck, breast, and flanks with blackish-brown; bill dark brown, pale at lower base; iris hazel; feet pale brown; claws darker, hind one long and little curved. Length 5·75. Female scarcely differs. After autumn moult, both sexes are darker above and have a pronounced brownish tinge below; migratory birds show this plumage on arriving on our South Coast in spring, and are also slightly smaller than residents.

Common throughout British Isles, frequenting chiefly rough and elevated districts, but also found in lower pasture-land; in winter gregarious and partially migratory. Nest: placed on ground in grass or heather; composed of dry grasses and lined with finer grass. Eggs; 4 or 5; greyish-white, mottled all over with chocolate-brown, and often with one or two black streaks; size ·80 by ·58; two broods are produced. Shrill but not unmusical song is usually uttered on the wing. Call note: a plaintive *peep*; flight-note: a low *gip, gip, gip*.

65. Anthus cervinus (Pall.). Red-Throated Pipit.

Hab. Northern regions of Old World, from N. Scandinavia to Kamschatka. In winter, southward to Africa and India.

Male: above of a more reddish-brown than in *A. pratensis*, and centres of feathers are blackish; above eye a distinct reddish-buff stripe; throat and breast pale chestnut, latter with a few small dusky spots; underparts pale buff, streaked on flanks with dusky-brown; bill as in *A. pratensis;* tarsi paler. Length nearly 6·00. Female: reddish tint confined to throat; breast and flanks more closely streaked. In winter both sexes almost lose reddish throat, and feathers of back have brownish-white margins.

A rare wanderer to W. Europe on migration. Two examples have occurred on South Coast in spring, viz., one near Brighton in 1884 (Zool. 1884, p. 192) and one in Kent in 1880 (Zool. 1884, p. 272); a third was shot near St. Leonard's in Nov. 1895 (Zool. 1896, p. 101). Another in the Bond collection was supposed to have been obtained in the Shetland Isles in 1854. It is said that in flight this species appears larger than *A. pratensis*, and has a longer and sharper call-note.

66. Anthus trivialis (Linn.). Tree-Pipit.

Hab. Europe (except extreme south), north to lat 70° in Norway; also Siberia east to the Yenesei. In winter southward to Africa.

Male: eye-streak buffish; feathers of upper parts dark brown in centre, with broad light buffish-brown margins; wings dark brown, all the feathers with pale outer margins; outer tail-feathers on each side chiefly white, next one tipped with white, rest dark brown; chin whitish; under parts pale yellowish buff, with a line of dark spots on each side of throat, and with bold elongated spots of dark brown on breast and flanks; centre of belly white; bill brown, paler at lower base; tarsi very pale brown. Length 6·00. Female: rather smaller and with smaller spots below. Hind claw is considerably shorter and more curved than in *A. pratensis*, and plumage is also paler in both spring and autumn.

Common throughout England (except extreme west) and South Scotland from April to September, but has not yet been identified in Ireland. Frequents timbered land or outskirts of woods. Nest: on ground concealed in a grass-tuft; composed of dry grass with a little moss, lined with fine grass and sometimes horse-hair. Eggs: 4 or 5; of several varieties, one being greyish-white closely freckled with dark red-brown, a second approaching eggs of Common Bunting, but of a rich reddish tint and smaller; commonest type is purple-grey with bold spots of dark brown; size ·83 by ·62. Song is commenced while the bird is perched on a tree, ascending as it soars upward, and dying away as it descends with outspread wings until it ceases abruptly as the singer regains his perch.

67. Anthus campestris (Linn.). TAWNY PIPIT.

Hab. Middle Europe, west to France, north to Holland and extreme south of Sweden, south to Mediterranean, east into Asia as far as N.W. India. In winter southward to Africa.

Male: above buffish-brown, inclining to greyish, and with dusky centres to feathers of crown and upper back; eye-stripe broad and whitish; wings dark brown with yellowish-buff margins to both quills and coverts; two outer tail-feathers on each side white on outer and portion of inner webs, rest of tail brown; chin and throa whitish, partly bordered by a very indistinct dusky line running from base of bill; breast pale tawny, with a few small spots of dark brown; belly and sides buffish-white; bill light brown above, yellow below; tarsi pale yellowish-brown; hind claw moderately long but curved. Length 6·50. Female scarcely differs; breast very little spotted.

A very scarce straggler to our shores during autumn migration. First recorded example was taken near Brighton in 1858, and a dozen or more have since been obtained there, together with one in the Scilly Isles (1868), and one in Yorkshire (1869).

68 Anthus richardi, Vieillot. RICHARD'S PIPIT.

Hab. Asia, north to lat. 58°, south to Turkestan and Mongolia. In winter straggling to most parts of Europe; usually wintering in India and China.

Male (in autumn): eye-stripe indistinct; feathers of head and mantle dusky-brown in centre, with broad rufous-buff margins; rump light brown; wings dark brown, with buffish-white outer edges to secondaries and primaries; wing-coverts with pale reddish-buff tips; two outer tail-feathers white on outer halves, rest of tail dusky-brown, with narrow buffish edges to the feathers; upper throat white, bordered each side by a line of dusky spots reaching to the breast, which latter is pale reddish-buff, spotted with dark brown; centre of belly white; bill dark brown above, pale below; tarsi long, pale brown; hind claw long and very little curved. Length fully 7·00. Female similar.

An irregular straggler to South Coast in autumn, a considerable number of occurrences having been reported. Northward has been recorded from Norfolk, Northumberland, Cumberland, Lancashire, Shropshire and Warwickshire, while at least one example has been obtained in Scotland.

69. **Anthus spipoletta (Linn.).** WATER-PIPIT.

Hab. Mountain regions of Central and Southern Europe; also Asia Minor and Western Asia. On migration strays to N.W. Europe; winters in Africa and India.

Adult (in autumn): above greyish-brown, browner on the rump; above eye and ear-coverts, a broad white streak; wings dark brown, coverts and secondaries with pale margins; outer tail-feather on each side quite white on its outer half, next two tipped with white, rest of tail, brown; chin and throat whitish; sides of neck and breast spotted with dark brown; belly buffish-white, flanks darker, slightly streaked with brown; bill and tarsi brown; hind claw moderately long and very little curved. Length nearly 6·50.

A very rare visitor; four undoubted specimens were taken in Sussex between 1864 and 1877, while it is almost certain that most of the recorded examples of *A. ludovicianus* (Harting, Handbook, p. 109), must have belonged to *A. spipoletta*.

70. **Anthus obscurus (Latham).** ROCK-PIPIT.

Hab. British Islands; also Channel Isles and northern coast of France.

Male: above olive-brown, with darker centres to the feathers; above eye an indistinct buffish-white line; edges of wing-coverts and secondaries slightly margined with pale brown; tail dark brown, outer feather on each side greyish on exterior half; under parts pale brownish-yellow, whiter on throat and centre of belly, and with an olive tinge on breast, which is spotted with dark brown, under parts being also slightly streaked; bill dark brown, paler at base; tarsi brown; claws dusky, hind one more curved than in *A. spipoletta*, and scarcely so long. Length 6·60. Plumage rather darker in winter than in spring. Female identical.

Tolerably common on all the rocky coasts of British Isles; to the flat shores of east side it is a winter visitor. Nest: placed in crevices of rocks, or often in a grass-tuft on a rock-ledge or declivity; composed of seaweed and grasses, with a lining of finer grass. Eggs: 4 or 5; light greenish-grey, closely mottled with dark olive-brown; there is also a reddish variety, but it is rare; size ·85 by ·62. It is often seen searching for small crustaceans, etc., among sea-refuse cast up on the beach; usually very silent, but when disturbed will fly a few yards uttering a shrill *peep, peep.*

71. Anthus rupestris, Nilss. SCANDINAVIAN ROCK-PIPIT.

Hab. Norway.

This brighter northern form is undoubtedly a valid species. The most important distinction is that, in breeding plumage, the throat and breast show a bright pinkish-buff tinge, which, however, is lost in winter, although at latter season both upper and under parts are distinctly paler than in *A. obscurus*. At all seasons, moreover, greater and lesser wing-coverts show more pronounced buffish-white tips, while the whitish stripe above eye is much more apparent. Length 6·20. According to the late E. T. Booth this form used to visit Sussex in some numbers during spring migration, and a few occurrences are also recorded from other parts of England, but at the present time it appears to be a rare straggler to our shores. A case of examples in breeding plumage may be seen in the Booth collection at Brighton.

Family Oriolidæ.

GENUS XXVIII. ORIOLUS, *Linnæus (1766).*

Bill moderately long, tolerably stout, conical and a little decurved, tip of upper mandible notched. Nostrils exposed. Wings rather long; 1st quill developed, 3rd longest. Tail moderate, rounded. Feet rather large, tarsus short.

72. Oriolus galbula, Linn. GOLDEN ORIOLE.

Hab. Europe, rare in extreme S.E., in north breeding to England on the west and Finland on the east; also North Africa. In winter southward to Africa.

Male: lores black; wings black, coverts and quills tipped and edged with whitish-yellow; tail-feathers black, all except central ones having broad yellow tips; rest of plumage above and below rich bright yellow; bill dark red; iris bright red; tarsi slate. Length 9·50. Female: wings dark brown, with grey edges and tips to the feathers; yellow of body duller and tinged with green; breast and belly striated with greyish brown; black lores absent, except in very old birds. Young: still duller than female.

A scarce summer visitor to southern half of England, particularly Cornwall, Devon, and Scilly Islands, to which it appears to be an annual visitor in spring. Nest has been found at intervals in nearly all the south-eastern counties; most recent instance being at Wicken, Cambridgeshire, in 1893. Nest: suspended from small branches of trees; deep and cup-shaped; woven of bark-strips, grass and wool. Eggs: 4 or 5, glossy white, with a few spots and blotches of dark red.

Family Laniidæ.

GENUS XXIX. LANIUS, *Linnæus (1766).*

Bill short, stout, compressed laterally; upper mandible with a pronounced tooth, and much hooked at tip;

base furnished with bristles. Wings moderate. Tail moderately long; rounded or graduated. Feet strong; claws curved.

73. Lanius excubitor, Linn. GREAT GREY SHRIKE.

Hab. Central Europe, west to North France, north and east to Scandinavia and St. Petersburg. Migrating from north in winter.

Male: sides of head black; a white streak above eye; upper plumage very light ash-grey; scapulars white; wings black, *with white bases to both primaries and secondaries,* both also being slightly tipped with white two centre tail-feathers black, others with broadening white tips, outer feathers each side being entirely white; lower plumage white; bill and tarsi brownish-black. Length 9·20. Female has breast marked with faint crescentic greyish bars.

A regular winter visitor to eastern side of Great Britain, but not common; of very rare occurrence in Ireland.

73a. L. excubitor major (Pallas). PALLAS'S GREY SHRIKE.

Hab. Northern Palæarctic region, from North Scandinavia and Central Russia, eastward through Siberia. Migrating southward in winter.

This race is distinguished by having white bases to the primaries *only*, bases of secondaries being black, and wing thus showing only one white spot or bar; it also seems to be very slightly larger than *L. excubitor.* Prof. Collett has proved that in Scandinavia it inter-breeds with the latter. In England it is of as frequent occurrence during winter as the typical form, and two Irish-taken Grey Shrikes in Dublin Museum both belong to this race.

74. Lanius minor, Gmelin. LESSER GREY SHRIKE.

Hab. Central and S.E. Europe, north to Prussia, west to S.E. France, casually north-west to England; also breeding in North Africa, Asia Minor, and eastward to Central Asia. Migrating southward in winter.

Male: frontal band, feathers round eye, and ear-coverts black; upper parts pale ash-grey, lightest on the rump; wings black, except basal-third of each primary which is white, forming a conspicuous patch; secondaries slightly tipped with white; outer tail-feather each side white, rest with white tip and base, and a black central portion, the white decreasing successively until two middle feathers are black; lower parts white, with a reddish tinge on breast and sides; bill and tarsi black. Length 8·50. Females and immature birds have very little black on head, and have under parts duller with a few faint crescentic bars.

Four occurrences are recorded, viz., one in Scilly Isles (1851), two at Yarmouth (1869 and 1875), and one near Plymouth (1876).

75. Lanius collurio, Linn. REDBACKED SHRIKE.

Hab. Europe (except extreme S.W.), north to South Scandinavia; also Asia Minor and Turkestan. In winter southward to Africa.

Male: crown, nape, and upper tail-coverts light grey; forehead, lores and ear-coverts black; back and lesser wing-coverts bright red-brown; wing-quills dusky brown, with reddish margins; two middle tail-feathers blackish, remainder white at basal-half, and with a narrow white tip, rest of feather being blackish; lower plumage cream-coloured, with a pink tinge; bill and tarsi black. Length 7·00. Female: above dull reddish-brown, with indistinct darker crescentic marks; below brownish-white, with more distinct greyish crescentic bars, these markings, however, especially on upper parts, become fewer as the bird ages, while a grey tinge appears on head, nape, and upper tail-coverts, and tail-feathers are edged with white. Young: browner, and more barred above and below than female.

Common from May to August throughout southern and midland counties, and many parts of Wales; rare in northern counties and more so in Scotland; in Ireland has occurred once near Belfast. Nest: placed in thick thorn bushes or tall hedges; height from four to ten feet; composed of moss, roots, grass-stalks, wool or feathers, lined with fine fibrous roots and hair; sometimes small twigs are used externally. Eggs: 4 to 6; yellowish-white, spotted with brown and grey, greenish-white spotted with olive-brown and grey, or reddish-white, spotted with red and grey; markings being chiefly around larger end and the grey being underlying spots; size ·90 by ·65. Two broods may be occasionally produced, as I have found half-fledged young on August 6th. Usual note, a loud harsh *chack, chack*, uttered while perched, and accompanied by both upward and lateral jerks of tail. Food: mice, small birds, bees, beetles, etc., which may be often found impaled on thorns, but *not* always near nest.

76. Lanius pomeranus, Sparrman. WOODCHAT.

Hab. Europe, north to Baltic and casually to British Isles; also Asia Minor, Persia, and N. Africa. In winter south to Africa.

Male: in front of eye, each side, a white streak, above which frontal band is black, as are also sides of head and a streak on each side of neck, bordering bright chestnut of nape and crown; upper back black; scapulars white; wings blackish, with white bases to primaries, and buffish-white tips to secondaries and greater coverts; rump light grey; outer tail-feather each side white, except a portion of inner web, rest of tail blackish, slightly tipped with white; lower parts yellowish-white. Length 7·25. Female: duller, back being largely rufous-brown, but scapulars are white.

Mr. O. V. Aplin has shown (Zool. 1892, p. 345—52) that not more than from 35 to 40 examples have been taken in England (from Northumberland to the Scilly Isles), and a few more seen, while it appears to have nested twice in the Isle of Wight, and is said to have done so in other localities; it is unknown in Scotland and Ireland.

Family Ampelidæ.

Genus XXX. AMPELIS, *Linnæus (1766)*

Bill short, strong, nearly straight, upper mandible decurved at tip and notched; gape wide, without bristles. Wings long; tail short, nearly even. Feathers of crown forming an erectile crest.

77. Ampelis garrulus, Linn. Waxwing.

Hab. Northern Palæarctic and Nearctic regions, from North Scandinavia to Alaska. Migrating southward in winter.

Male: general plumage light brown, with a greyish tinge, becoming light grey on lower back and belly; sides of crest and the forehead chestnut-brown; below latter a black frontal band, running backward to lores and circumocular regions; throat also black; greater wing-coverts black, tipped with white; quills black; primaries barred alternately with white and yellow; secondaries spotted with white at end and with peculiar tips like red sealing-wax to shafts; tail blackish, with a broad yellow terminal band, and, in many old birds, with the red wax-like tips to shafts; lower tail-coverts reddish-brown; bill and tarsi black. Length about 7·50. Female: duller generally, and with few of the wax-like tips to secondaries.

An almost annual winter visitor to British Isles; on eastern side occasionally appearing in some numbers. Feeds at this season upon berries of various kinds.

Family Muscicapidæ.

GENUS XXXI. MUSCICAPA, *Linnæus (1766).*

Bill moderate, base wide and depressed, tip compressed and slightly decurved; gape furnished with bristles; wings rather long, pointed; feet small.

78. Muscicapa grisola, Linn. SPOTTED FLYCATCHER.

Hab. Europe, north to lat. 70°; also N. Africa, Palestine, and Asia, eastward to Lake Baikal. In winter migrating to Africa and India.

Male: cap pale brown, with a dark brown streak in centre of each feather; back and lesser wing-coverts dull brown, tinged with olive; wings dusky-brown; greater coverts and secondaries having paler edges; tail dark brown; throat, breast and sides buffish-white, with small streaks of brown; chin, centre of abdomen, and under tail-coverts, pure white; bill dark brown, yellowish at lower base; feet brownish-black. Length 5·75. Female: duller and rather less spotted on under parts. Young: head and nape mottled with white and blue-grey, and mantle with various shades of brown; under parts white, tinged with pale brown on breast and flanks, but not spotted.

A common visitor to British Isles, arriving early in May and leaving at end of September. Frequents woods, parks, gardens, shady lanes, etc. Nest: commonly placed on small branches against trunk of a tree, or on a horizontal limb at a height of from five to twenty feet; sometimes in ivy or creepers, or in holes in trees or walls; I have found it built like a chaffinch's in the fork of a bush; it is composed of moss, dry grass, hair and feathers, lined with hair, feathers, fine strips of inner bark, and often small pieces of cast snake's-skin. Eggs: 3 to 5; dull white, tinged with pale bluish-green and spotted or blotched with dull rusty-red, with underlying spots of pale purple; size, ·75 by ·56. Food consists of insects taken on the wing; flight is noiseless and circling, the bird returning constantly to the same perch. Call note, a weak shrill chirp, but seldom heard.

79. Muscicapa atricapilla, Linn. PIED FLYCATCHER.

Hab. Europe, north to lat. 70° in Norway, but local in many parts, in winter migrating to Africa.

Male: forehead white; rump dusky-grey; remaining upper plumage blackish; secondaries with broad white outer margins forming a conspicuous patch on the wing; two outer tail-feathers each side slightly margined with white; lower plumage white; bill and tarsi black. Length 5·10. In autumn upper parts are brownish-black. Female: upper parts dull brown; white on forehead and wings, tinged with brown; under parts brownish-white. Young birds, after autumn moult, resemble female.

Locally common from May to September in North of England and in Wales but seldom breeding in the south or east; in Scotland breeds sparingly in the south; in Ireland five stragglers have been taken. Nest: placed in holes in trees or walls; loosely composed of dry grass, moss, fine roots, feathers and hair. Eggs: 5 to 8; pale blue; like Redstarts but paler, more oval, and slightly smaller;

size ·69 by ·52. Like *M. grisola* it feeds on insects but is not so much addicted to taking them on the wing; the male utters a tolerably loud and pleasing warble.

80. Muscicapa parva, Bech. REDBREASTED FLYCATCHER.

Hab. Eastern half of Europe, north to St. Petersburg, south to Black Sea; also Asia, east to lake Baikal. In winter migrating to Africa and India, and casually to W. Europe.

Male: upper parts dull brown, becoming greyish on head and nape, and pale grey on cheeks; basal half of each tail-feather (except two middle ones) white, rest of tail dusky-brown; throat and breast, light orange-red; abdomen buffish-white; bill and tarsi dark brown. Length about 4·80. Female: orange-red is duller and restricted to the throat. Young: like female but wing-feathers are tipped and margined with buff and mantle spotted with same.

Nine occurrences only are recorded for British Isles.

Family Hirundinidæ.

GENUS XXXII. HIRUNDO, *Linnæus (1766).*

Bill short, weak, depressed, and very wide at base. Wings long, pointed, with nine primaries only; tail of twelve feathers, much forked; outer feathers, very long; tarsus without feathers; feet small.

81. Hirundo rustica, Linn. SWALLOW.

Hab. Europe, north to lat. 70° in Norway, and northern Asia, east to lake Baikal; also North Africa. In winter to S. Africa and India.

Male: upper parts metallic blackish-blue; forehead, throat, and under tail-coverts, chestnut red; upper breast, blackish-blue; abdomen, light buff; wing-quills and tail black with green reflections, all the tail-feathers, except the short middle ones, having basal portion of inner web white; bill and feet black. Length to tip of tail 7·75. Female: tail shorter, and under parts paler. Young: duller above, and still paler below, while the tail is much shorter than in adults.

Common from mid-April to mid-October, except in extreme north of Scotland and West Coast of Ireland, where it is scarce and local. Nest: placed on rafters in barns or under bridges, in angles of porches, etc.; composed of mud, mixed with grass, and lined with fine grass and feathers. Eggs: 4 or 5; white spotted and blotched with reddish-brown and with underlying grey marks; size ·82 by ·55. Two broods are produced, young being fed entirely upon small-winged insects, which also constitute food of the parents. In spring the male utters a weak but sweet "twittering" song as he courses over the meadows.

GENUS XXXIII. CHELIDON, *F. Boie (1822)*.

Differs from *Hirundo* in having tarsi and toes clothed with short feathers; tail is forked, but outer feathers scarcely exceed the next.

82. Chelidon urbica, (Linn.). MARTIN.

Hab. Europe, north in Scandinavia to lat. 70°; also N. Africa, Asia Minor, and eastward to N.W. India. Migrates southward in winter.

Male: upper parts glossy black, except rump which is white; whole under parts white; bill black; feet clothed with short white feathers. Length 5.25. Female: identical. Young: blackish-brown above, with dull white rump and lower parts; tail less forked.

Common throughout British Isles from the end of April to mid-October, except in North Scotland, where it is rather scarce. Nest, placed beneath the eaves of buildings, under bridges, or against the face of a cliff; never open at the top like a swallow's, an entrance hole being made in one side; constructed of mud pellets, and lined with grass and feathers; usually much infested with the bird's parasites. Eggs: 4 or 5; immaculate white; size, ·80 by ·52. Two or three broods are produced. Food: winged insects.

GENUS XXXIV. COTILE, *F. Boie (1822).*

Differs little from *Chelidon* except in having feet naked, save for a few feathers above the hind toe.

83. Cotile riparia (Linn). SAND-MARTIN.

Hab. Circumpolar, breeding in Europe northward to lat. 70° in Norway. Migrating southward in winter.

Male: above mouse-brown, wings and tail blackish; below white, excepting a pale brown band across breast; bill black; tarsi reddish-brown. Length nearly 4·75; tail moderate, distinctly forked. Female differs little. Young: darker below, and feathers of upper parts are tipped with buffish-white.

Locally common throughout British Isles from end of March to late in September; Mr. Ussher says that in Ireland it breeds more commonly than the House-Martin. Breeds in colonies in sandstone cliffs, railway cuttings, etc.; nesting hole is bored by the bird and penetrates about two feet with a slight ascension; nest consists of a little grass and feathers. Eggs: 4 to 6; pure white; size ·72 by ·50. Two broods are produced; the birds are rather bold and will pursue birds of other species which approach their nests, uttering a low and rather harsh alarm-note; male has also a weak twittering song.

Family Fringillidæ.

Sub-Family Fringillinæ.

GENUS XXXV. CARDUELIS, *Brisson (1760).*

Bill moderately short, nearly conical, slightly compressed. Nostrils concealed by recurved feathers. Wings rather long, 1st primary obsolete, 2nd longest. Tail moderate, slightly forked. Tarsus rather short; claws moderate, not much curved.

84. Carduelis elegans, Stephens. GOLDFINCH.

Hab. Europe, north to Southern Scandinavia; also N. Africa, Madeira and Persia.

Male: lores black; forehead, forepart of cheeks, and upper throat crimson; remainder of cheeks and lower throat white, bordered above and behind by the black of crown and sides of nape, centre of latter being whitish; back brown; upper tail-coverts chiefly white; wings black, quills banded with bright yellow and tipped with white; tail black, tipped with dull white, three outer feathers each side having oblong white central patches; breast and sides brownish-yellow; belly white; bill yellowish-white, tip blackish; tarsi pale brownish-yellow. Length 5·00. Female rather duller. Young: red and black of head are replaced by greyish-brown.

Breeds sparingly in every part of British Isles, although constant persecution has caused it to become rare in many localities. Nest: placed in hedges or in the fork of a tree in orchards and gardens; somewhat slighter than the Chaffinch's and composed of moss, spiders' cocoons and grass-stalks, lined with plant-down, hair and feathers. Eggs: 4 or 5; pale greyish-blue, rather glossy, and spotted and streaked sparingly with dark red brown and faint purplish-red; size ·68 by ·50. Two broods are produced. Food consists chiefly of seeds of thistle, dock, etc., but young are fed upon insects and small caterpillars. A portion of the birds which breed here leave us in winter, but small parties remain and frequent waste lands, etc.

GENUS XXXVI. CHRYSOMITRIS, *Boie (1828).*

Bill as in *Carduelis* but shorter; wings and tail also similar; tarsi more slender; claws sharp and curved. Size small; general tint of plumage greenish-yellow.

85. Chrysomitris spinus (Linn.). SISKIN.

Hab. Palæarctic Region; in Europe breeding from within Arctic Circle to Central Europe.

Male: crown, lores and chin black; cheeks and ear-coverts olive-green, bordered above by a broad stripe of yellow; nape, mantle and lesser wing-coverts olive-green, streaked with blackish; rump yellow, wings blackish, greater-coverts tipped with yellow, and quills with yellow margins and bases; two middle tail-feathers and tips of remainder blackish, rest of tail yellow; lower throat and breast greenish-yellow; sides yellowish with dusky streaks; centre of belly white; bill yellowish-brown; tarsi brown. Length 4·75. Female: crown merely streaked with black, and yellow of plumage is much duller.

Breeds in Cumberland, and perhaps other northern counties, and generally throughout East Scotland, but more rarely in the west; in Ireland breeds locally in most of counties on east side. To greater part of England and Wales a rather common visitor in winter or on migration, but very rarely nesting. Frequents chiefly plantations of coniferous trees. Eggs: similar to those of Goldfinch but of a more greenish-blue tint; size ·65 by ·48.

GENUS XXXVII. SERINUS, *Koch (1816).*

Bill short, stout, conical, upper mandible very slightly longer than lower. Wings with 1st quill minute, 3rd longest. Tarsus moderate; claws small.

86. **Serinus hortulanus**, *Koch.* SERIN.

Hab. Southern half of Europe, north to the Rhine; also N. Africa and Asia Minor. Casually to British Isles.

Male: above greenish-olive streaked with dark brown, except rump, forehead and a streak above eye which are yellow; tail and primaries dark brown, with yellow margins; secondaries and greater coverts edged with buffish-white; throat and breast yellow; centre of belly white; sides broadly striped with dark brown; bill dark brown; tarsi light brown. Length 4·50. Female: yellow tints very much duller.

A rare straggler on migration; seven or eight have been taken in Sussex; has also occurred in Somerset, Norfolk, near London, and once in Ireland (1893).

GENUS XXXVIII. LIGURINUS, *Koch (1816).*

Bill short, stout, conical, tip of upper mandible compressed and very indistinctly notched. Wings fairly long, 1st quill absolute. Tail very little forked.

87. **Ligurinus chloris** (Linn.). GREENFINCH.

Hab. Europe (except Spanish Peninsula), north to southern Scandinavia; also Turkestan.

Male: above each eye a yellow stripe; lores blackish; forehead and rump greenish-yellow; crown, nape and mantle olive-green; wing-quills greyish-brown, primaries having outer webs chiefly yellow; basal halves of tail-feathers, except two central ones, yellow, terminal portions dusky brown with paler margins; whole lower parts dull pale yellow; bill dull yellowish-white, darker at tip; tarsi pale brown. Length nearly 6·00. Female: slightly smaller; plumage less yellow. Young: general tint is brown, tinged with yellow and somewhat striated.

A common resident, frequenting chiefly arable land and placing nest in hedgerows and bushes at a height of 6 or 8 feet; constructed of roots, twigs, moss and wool, lined with wool, feathers and hair. Eggs: 4 to 6; whitish or greenish-white, spotted and blotched with red-brown, light red and grey; size ·82 by ·57. Two broods are produced, first eggs being laid at end of April. Call-note: a monotonous, plaintive *cree-e-e-e*, uttered while perched on a hedge or tree. The weak tremulous song may be syllabled as *tittle-tittle-tee, ter-ter-tee.*

GENUS XXXIX. COCCOTHRAUSTES, *Brisson (1760).*

Bill rather short, conical, very stout; mandibles almost equal, upper one rounded. Wings with 1st primary obsolete, 3rd longest, inner ones curved outward and with jagged tips. Tail short, even. Tarsus short, stout; claws strong, somewhat curved.

88. Coccothraustes vulgaris, Pallas. HAWFINCH.

Hab. Europe, south of lat. 60° N., but very local; also Asia Minor and N. Africa.

Male: crown and cheeks orange-brown; lores, feathers at base of bill, and throat black; nape ash-grey; mantle brown; wings blackish, browner on secondaries; greater coverts chiefly white; primaries with glossy blackish-blue tips and patches of white on inner webs; upper tail-coverts brownish-orange, reaching beyond middle of tail which is white on terminal half and black at base; under parts pale rufous-brown; bill bluish, blacker at tip; tarsi very pale brown. Length 7·00. Female duller. Young:

paler and somewhat mottled both above and below; black throat absent until autumn moult.

Breeds commonly in Middlesex, Essex, Kent and Surrey; sparingly also in most other south-eastern and midland counties, north perhaps to Yorkshire, where it has nested twice recently. To south-western and northern counties and Wales it is a winter visitor; has occurred at same season in Scotland, and with greater frequency on east side of Ireland. Frequents woodlands, building nest in tall thick hawthorns or hollies, the lower branches of oak trees or in fruit trees in orchards; it is cup-shaped, but shallow, and very lightly made of fine twigs and roots, lined with bark-strips and horse-hair. Eggs: 4 or 5; light olive-grey, boldly spotted with blackish-olive and with some thick streaks of grey; size ·95 by ·70. Call-note: a peculiar prolonged whistle, but note when intruders are near is *pit*, *pit*, *pit*, often repeated. Food: berries, peas, seeds of hornbeam and kernels of fruit-stones, which are broken in its strong bill.

GENUS XL. PASSER, *Brisson (1760)*.

Bill short, nearly conical, but both upper and lower mandibles somewhat arched, and former slightly larger than latter. Wings with 1st quill small, 3rd or 4th longest. Tarsus moderately long, stout; claws short.

89. Passer domesticus (Linn.). HOUSE-SPARROW

Hab. Europe (except Italy and islands of Mediterranean), north to Arctic Circle; also Siberia.

Male: crown and nape grey, bordered with chestnut; above eye a slight streak of white; lores black; mantle chestnut, streaked with black; lower back grey; median wing-coverts broadly tipped with white; quills and tail brown; chin and throat black;

cheeks, sides of neck, and belly dirty white; bill and claws blackish; tarsi brown. Length 6·00. In winter bill is brown and plumage duller. Female: duller and browner; lacks grey head and black throat.

Common and resident both in towns and country; gregarious in winter. Nest: placed in holes of all kinds, and in trees or hedges; nests of House and Sand Martins are often usurped. Eggs: 5 or 6; greyish or greenish-white, spotted very variably with dusky-brown, olive and grey; size ·85 by ·60.

90. Passer montanus (Linn.). TREE-SPARROW.

Hab. Palæarctic region; in Europe north to lat. 70° (Norway), but rare in extreme south.

Male: crown, nape and lesser wing-coverts soft reddish-brown; lore, and a streak under and behind eye black; cheek and side of neck white, with a noticeable black patch in centre; upper parts as in *P. domesticus*, but both median and greater wing-coverts tipped with white, forming a double bar; throat and upper breast black; belly greyish white; bill blackish; iris hazel; tarsi light brown. Length 5·75. Female identical; young also scarcely differ, even in nestling plumage.

Generally distributed, but scarcer than *P. domesticus*, and unknown in towns; breeds very commonly in Middlesex, Cambridgeshire, Essex, Kent, and regularly over midland and eastern counties; rare in south-west Wales, counties north of Nottinghamshire and Scotland. In Ireland, not uncommon in Co. Dublin, but apparently overlooked elsewhere. Nest: in narrow-apertured holes in pollard willows and oaks, sometimes in thatched roofs, often in holes or in crevices of sea-cliffs; composed of dry grass and a quantity of feathers. Eggs: 4 to 6; readily distinguished from House-Sparrow's; white, closely freckled with rich chocolate-brown, but in nearly every clutch is one "light" egg, sparingly marked with greyish-brown; size ·78 by ·55. Young, like those of House-Sparrow, are fed upon small caterpillars, but old birds feed largely on seeds; in winter small flocks frequent stubbles. Call note: a sharp *chuck, chuck*, less loud and aggressive than House-Sparrow's.

GENUS XLI. FRINGILLA, *Linnæus (1766)*.

Bill rather longer than in *Passer*, mandibles nearly equal. Wings with 1st quill obsolete, 2nd rather short, 3rd or 4th longest. Tail somewhat forked.

91. Fringilla cælebs, Linn. CHAFFINCH.

Hab. Western Palæarctic region; in Europe north to lat. 70° (Norway).

Male: forehead black; crown and nape bluish-grey; mantle reddish-brown; lower back yellowish-green; lesser wing-coverts white, the greater black tipped with yellowish-white; quills dark brown, with pale greenish outer edges; two middle tail-feathers slate, rest black with large white patches on two outermost feathers each side; sides of head and lower plumage light reddish-brown; bill dusky; iris hazel; tarsi brown. Length 6·00. Female: slightly smaller; head and mantle dull brown; white on wings less noticeable; under parts brownish-white, with a rufous tinge on breast. Young resemble female.

Common everywhere; partially migratory in winter when numbers also arrive from Continent; sexes separate at this season, many flocks consisting entirely of females, while old males are often seen singly. Nest: placed in bushes, hedges or lower branches of trees; neatly composed of moss, wool and hair, lined with hair and feathers; lichens, pieces of decayed wood, or spider's cocoons are often stuck on outside to harmonize with the surroundings. Eggs: 4 to 6; pale greenish or greyish-blue, spotted and clouded with reddish-brown; sometimes whole surface is suffused with brownish-red, while unmarked bluish eggs are not rare; size ·78 by ·56 Call-note: a sharp *pink*, *pink;* has also a sweet and musical song.

92. Fringilla montifringilla, Linn. BRAMBLING.

Hab. Northern Palæarctic region, north of lat. 60° in Europe and 50° in Asia. Migrates southward in winter.

Male in spring: feathers of head, nape and mantle bluish-black, with more or less concealed whitish bases; rump white, slightly mottled with black; lesser wing-coverts buffish-orange, with white tips; greater coverts black, broadly tipped with white; quills blackish, with narrow white outer edges; tail black, outer feather each side partly white at base; throat and breast pale brownish-red; belly whitish; sides mottled with black; under wing-coverts yellow; bill blackish; tarsi brown. Length 6·00. In winter feathers of head and mantle have broad rufous margins, and bill is yellow, except for a dusky tip. Female: duller; head and mantle are dark brown.

A common winter visitor to Scotland and fairly so to east of England; more rarely to west and Ireland. The late E. T. Booth found it nesting in Perthshire in 1866.

GENUS XLII. CANNABINA, *Boie (1828).*

Bill moderate, almost conical, tip pointed and sharp. Wings rather long, 1st primary absolete, 2nd or 3rd longest. Tarsus moderate; hind claw rather long, curved.

93. Cannabina cannabina (Linn.). LINNET.

Hab. Europe, north to South Scandinavia; also N.W. Africa and Madeira.

Male: fore-part of head crimson; nape and neck ash-brown; back and lesser wing-coverts reddish-brown; primaries blackish with narrow white outer edges; upper tail-coverts dusky brown, margined with buffish-white; tail blackish, all except two middle feathers edged on both webs with white; throat, breast and sides buffish-white, striped with ash-brown, breast being suffused with bright crimson; belly whitish; bill dusky at tip, paler at base; tarsi brown. Length 5·50. In winter crimson of head and breast is lost. Female: browner and slightly smaller; more streaked above and below with dark brown and crimson is lacking. Young: like female.

Common, except in extreme north of Scotland. Frequents furze-covered commons or waste lands, nest being built in small bushes or furze; composed of roots, moss, wool, etc., lined with hair, fibrous roots, etc. Eggs: 4 to 6; greyish-white, spotted and streaked around larger end with reddish-brown; size ·72 by ·52. Food consists of seeds of various kinds; in autumn small parties frequent the stubbles. Song, which may be heard both in spring and autumn, is varied and sweet, but not very loud; the males will often sing aggressively against one another; I have heard them sing on the wing.

94. Cannabina linaria (Linn.). MEALY REDPOLL.

Hab. Northern Palæarctic region; in Europe north of about lat. 60°. Southward in winter to middle Europe.

Male: forehead dark crimson; lores and upper throat black; feathers of upper parts dark brown, with greyish or buffish-white margins; rump-feathers almost entirely pinkish-white; tail-feathers dark brown, with narrow paler edges; median and greater wing-coverts tipped with dull white; throat and breast suffused with carmine-red; sides streaked with dark brown: centre of belly whitish;

bill yellowish, darker at tip ; tarsi dark brown. Length 5·25. In winter upper parts are paler, mantle-feathers having broad dull white margins, and red of breast has disappeared. Female : slightly smaller, duller, and without red on breast.

A not uncommon winter visitor to Scotland and north and east of England, rarer in west, while it has been recorded twice from Ireland (Kildare 1876 ; Mayo 1893).

94a. C. linaria holbœlli (Brehm). HOLBŒLL'S REDPOLL.

Hab. North Europe from Scandinavia to E. Siberia.

A rather larger race ; Dr. Sharpe says that the bill measures ·50 instead of ·40 as in the typical form. Two examples from Norfolk are in the Natural History Museum.

95. Cannabina rufescens (Vieill.). LESSER REDPOLL.

Hab. Central Europe and British Islands.

Male : upper parts darker than in *A. linaria*, feathers being dusky brown in centre, with warm buffish margins ; rump is scarcely paler than mantle, but with a decided carmine tinge ; wing-coverts tipped with warm buff instead of white ; under parts slightly darker ; considerably smaller, also, than *A. linaria*, length being only about 4·25. In winter red of rump and breast is nearly absent. Female : forehead red, but not breast or rump. Young lack even the red of forehead.

Common over British Isles in winter, but scarcer in spring ; breeds in fair numbers from Norfolk and Nottinghamshire to the north of Scotland, also sparingly in Middlesex and other south-eastern and midland counties, and in most counties of Ireland. Nest : in tall hedges, saplings, or on small branches against a tree-trunk; rather small; composed of small twigs, grass, moss, etc., lined with willow-down and a few feathers. Eggs: 4 to 6; light bluish-green, spotted with pale reddish-brown; size ·61 by ·46. In winter roves in small parties, usually feeding in the upper branches of birches, etc.; alarm note is a sharp, but not loud, *bee-ing*.

96. Cannabina hornemanni (Holbœll).
GREENLAND REDPOLL.

Hab. Greenland and Eastern North America.

A large-sized Mealy Redpoll, considered by Dr. Sharpe to be a sub-species of *A. exilipes*, but American ornithologists make *A. hornemanni* the type, and *exilipes* its sub-species. A single specimen was shot in Durham in 1855 as recorded by the late John Hancock.

97. Cannabina flavirostris (Linn.). TWITE.

Hab. N.W. Europe, north in Scandinavia to lat. 70°.

Male: feathers of crown, nape, mantle, and upper tail-coverts dark brown with paler margins; rump suffused with crimson; wings dark brown, greater coverts and most of quills having dull white outer margins; tail dusky-brown with dull white margins to the three outer feathers on each side; lores, chin and throat light rufous; breast and sides pale buff, streaked with dark brown; centre of belly white; bill light yellow; iris hazel; tarsi dusky-brown. Length 5·25. Tail is distinctly longer than in the Redpolls, and there is no crimson on head or breast. Female: lacks crimson of rump; greater coverts are margined with pale brownish instead of white, and bill is brown at tip and yellowish at base.

A winter visitor to southern half of England, but breeding commonly from Yorkshire northward to the Shetlands; also in most parts of Ireland. Frequents moorlands and mountainous districts, placing its nest in heather or in a grass-tuft on rocky ledges. Eggs: 4 to 6; light bluish-green, spotted with reddish-brown and streaked with paler red; size ·70 by ·50. The call-note is imitated in the trivial name; male has also a slight song.

Sub-Family Loxiinæ.

Genus XLIII. PYRRHULA, *Brisson (1760).*

Bill short, stout, very strong, wide at base, bulging at sides, upper mandible longer than lower and slightly decurved at tip. Nostrils concealed by small feathers. Wings moderate, 1st primary obsolete. Tail moderate.

98. Pyrrhula europæa, Vieill. BULLFINCH.

Hab. Western and Central Europe, northward to Baltic and British Isles.

Male: cap and chin bluish-black; nape and mantle pale ash-grey; greater wing-coverts black, broadly tipped with greyish-white; quills and tail black; rump white; below bright salmon-red; under tail-coverts white; bill black; tarsi dark brown. Length 6·00. Female: cap duller; mantle greyish-brown; under parts dirty brown with a slight reddish tinge. Young: like female but black of head is absent.

Common and generally distributed. Frequents woods, copses and thicket-covered commons. Nest: in tall thorn bushes or hedges; cup-shaped but shallow; composed of fine twigs, neatly lined with fine fibrous roots and hair. Eggs: 4 to 6; light greenish-blue, spotted and streaked around larger end with dark purplish-brown and lilac-grey; pure white eggs, usually with normal markings, are rather common in Surrey; size ·75 by ·55. Call-note; a short, plaintive whistle often repeated; song, heard in early spring, is weak and soft.

98a. P. europæa major (Brehm). NORTHERN BULLFINCH.

Hab. Northern Europe, west to Norway, south to East Prussia and Poland; also Siberia.

Distinctly larger than typical bird, also having colours more brilliant and greater wing-coverts broadly tipped with pure white instead of greyish-white.

Two examples shot in Yorkshire in November, 1894, were exhibited at a meeting of the Zoological Society of London in November, 1895.

GENUS XLIV. CARPODACUS, *Kaup (1829).*

99. Carpodacus erythrinus (Pall.). ROSY BULLFINCH.

Hab. Palæarctic region; in Europe westward to Finland and Poland. A straggler to W. Europe.

One example was obtained near Brighton in September, 1869; another at Caen Wood, Middlesex, in October, 1870.

GENUS XLV. PINICOLA, *Vieillot (1807).*

100. Pinicola enucleator (Linn.). PINE-GROSBEAK.

Hab. Northern Palæarctic region, north to lat. 70° in Norway; also sub-Arctic America.

More than two dozen occurrences in all have been recorded from various localities, ranging from Kent to North Scotland, but subsequent examination has shown that not more than four or five of these are really reliable.

GENUS XLVI. LOXIA, *Linnæus (1766).*

Bill moderately long, strong, stout at base, both mandibles tapering towards tips, which are much curved and cross one another. Wings rather long, pointed, 1st primary minute. Tarsus short; claws tolerably large, rather curved.

101. Loxia curvirostra, Linn. CROSSBILL.

Hab. Palæarctic region, north to within Arctic Circle. Partially migratory in winter.

Male: greyish-brown, suffused more or less generally with light crimson; tail and wings dark brown, with pale margins to wing-coverts; bill and tarsi dark brown. Length 6·00. Young males are dull greenish-brown, palest below, with darker striations, and with a yellow tinge on the rump. Female: suffused with bright greenish yellow instead of red.

Fairly frequent during winter in England, and has also nested occasionally in nearly every county, while it breeds regularly in many parts of Scotland; in Ireland breeds irregularly in the south and east. Nest: on horizontal branches of firs at variable heights; composed of twigs, dry grasses, wool, etc., lined with finer grass and hair. Eggs: 4 or 5; bluish or greyish-white, sparingly spotted with two shades of reddish-brown; larger than Greenfinch's; size ·90 by ·68; usually laid early in March. Food: seeds obtained by prizing open fir-cones, also berries and apple-pips, with some insects in summer.

101a. L. curvirostra pityopsittacus (Bech.). PARROT CROSSBILL.

Hab. Pine forests of Scandinavia and Northern Russia; migrating southward in winter.

Formerly considered a valid species, but recent workers scarcely admit its claim to sub-specific rank. Distinguished from typical birds by its rather larger size and stouter bill.

Examples have been taken at intervals in most of our eastern and southern counties, and three or four times in Scotland; in Ireland it appeared rather commonly in winter of 1890-91 (Zool. 1891, p. 112), but only two examples had been obtained previously.

102. Loxia bifasciata (Brehm). TWO-BARRED CROSSBILL.

Hab. Northern Russia and Siberia. In winter migrating to Central and Western Europe.

Male: above blackish-slate suffused with scarlet, which largely conceals darker ground tint; greater and middle wing coverts with broad reddish-white edges and tips, forming two bars; secondaries also narrowly tipped with reddish-white; throat and breast rich scarlet; belly nearly white; bill brownish-yellow; tarsi brown. Length 5·80. Female: red tint replaced to a lesser extent by light yellow; both upper and under parts are also streaked with dark brown. Young: like female but with very little yellow, except on rump, and white wing-bands less distinct.

A rare straggler. Has been recorded at intervals from nearly all our southern and eastern counties, and twice from Ireland.

102a. L. bifasciata leucoptera (Gmelin). WHITE-WINGED CROSSBILL.

Hab. Northern North America, from Alaska to Labrador and Newfoundland.

Most British authorities now consider this form to be only entitled to sub-specific rank. Mr. Saunders says "the only difference of any moment between the European and American forms consists in the darker scapulars of the latter: to which I may add that the red in the male has a pinker tint, and the bill in both sexes is weaker." Three examples have been taken in England.

Sub-Family Emberizinæ.

GENUS XLVII. EMBERIZA, *Linnæus (1766).*

Bill short, conical; palate usually having a small bony projection. Tail moderately long, a little forked. Wings with 1st primary absolute. Tarsus moderate, covered with scales in front and laterally with a single plate, sharply ridged behind; hind claws moderate, curved.

103. Emberiza melanocephala, Scopoli.
BLACK-HEADED BUNTING.

Hab. S.E. Europe, Asia Minor, Palestine and Persia; wintering in Northern India. Straggles on migration to rest of Europe.

Three occurrences are on record, *i.e.*, one near Brighton (1868), a second in Nottinghamshire (1884), and a third in Dunfermline (1886).

104. Emberiza miliaria, Linn. CORN-BUNTING.

Hab. Western Palæarctic region, north to Baltic and South Norway. Partially migratory in winter.

Male: feathers above light brown, with narrow dusky centres; wings dusky-brown, with pale buff margins to greater and lesser coverts; tail-feathers dark brown, with paler margins; through and above eye a pale buff stripe; below buffish-white, throat bordered each side by a line of dark brown spots, breast also thickly spotted, and flanks streaked; bill brownish-yellow, darker on culmen; iris dark hazel; tarsi brownish-yellow. Length 7·00. Female: identical. Young have a pronounced buff tint below, and feathers of wings and mantle are much margined with same.

Generally but sparingly distributed, frequenting chiefly arable land. Nest: among corn, meadow-grass or clover, or in rank herbage on a bank; composed of moss, roots and grass, lined with finer grass. Eggs: 3 to 5; vinous-white, heavily blotched and streaked with deep chocolate-brown and faint purple; size ·95 by ·70. Call note: a loud clear *clink*, *clink*, uttered on the wing as well as while perched.

105. Emberiza citrinella, Linn. YELLOW HAMMER.

Hab. Europe, northward to lat. 70° in Norway; also Siberia, eastward to the Yenesei.

Male: cap and sides of head pale yellow; feathers of mantle and lesser wing-coverts reddish-brown, with dusky central streaks; wing-quills dusky, edged with greenish-yellow; rump and upper tail-coverts light chestnut; tail-feathers dusky, two outer on each side having a large patch of white on inner webs; below pale yellow, suffused with chestnut on breast and sides; flanks and under tail-coverts streaked with dark reddish-brown; bill slate-brown above, bluish below; iris brown; tarsi pale brown; claws darker. Length 6·60. Female: yellow and chestnut tints much less evident; head, breast and sides streaked with reddish-brown.

Very common and resident; chiefly, however, frequenting arable land. Nest: near the ground in low bushes, brambles, rank vegetation, etc., or even *in* the ground on open downs. Eggs; 3 to 5; varying in ground from bluish-white to light dingy brown, with spots, marblings or streaks and long lines of purplish-brown; size ·86 by ·65. Song: *tic-tic-tic-tic, tee-e-eeze*, ascending quickly, the termination prolonged and dying away more slowly.

106. Emberiza cirlus, Linn. CIRL BUNTING.

Hab. Western Europe, north to England, also Southern Europe, Asia Minor and Algeria.

Male: crown and nape olive with black streaks; above eye a pale yellow stripe; lores and ear-coverts black, forming a black band through eye, below which is another stripe of pale yellow; mantle chestnut-brown; *lesser wing-coverts greenish grey*; secondaries broadly edged with chestnut; rump and upper tail-coverts olive green, with dusky streaks; chin and throat black, below which is a band of pale yellow extending to ear-coverts; breast and sides olive-grey, streaked with chestnut; belly pale yellow; bill and tarsi as in Yellow Bunting; iris hazel. Length 6·00. Female: lacks black throat; chestnut and yellow replaced by duller tints.

Breeds tolerably commonly in the southern counties from Kent to Devon, also sparingly along the northern side of the Thames; has straggled casually to Scotland and Wales, but not to Ireland. Nest: in furze or at side of a hedge, often in a little bush just within a meadow. Eggs: 3 to 5; bluish-white with bold spots and short thick streaks of brownish-black and faint purplish-grey; size ·85 by ·62.

107. E. hortulana, Linn. ORTOLAN BUNTING.

Hab. Western Palæarctic region, in summer north to Trondhjem in Norway, but only casually to Britain.

A rare and irregular visitor in spring or autumn to eastern and southern coasts of England; two examples have been taken in Scotland and one in Ireland.

108. Emberiza rustica, Pallas. RUSTIC BUNTING.

Hab. Northern Russia and Siberia, migrating westward and southward in winter.

Three examples have been taken, *i.e.*. near Brighton (1867), Yorkshire (1881), and Elstree Reservoir, near London (1882).

109. Emberiza pusilla, Pallas. LITTLE BUNTING.

Hab. North-eastern Europe and Siberia. Migrating in winter to S. Asia and casually to Europe.

A single example was taken near Brighton in Nov., 1864.

110. Emberiza schœniclus, Linn. REED-BUNTING.

Hab. Palæarctic region, north to lat. 70° in Norway. In winter partially migratory.

Male: entire head black, bordered behind by a white collar and below by a white stripe running from base of bill and joining collar; throat and centre of breast black, sides of latter white; sides of abdomen light brown, with dusky streaks, centre white; feathers of mantle and wing-coverts reddish-brown, with dusky centres; wing quills and tail dusky, two outer feathers on each side of latter with white patches on inner webs; bill dark brown above, paler below; iris hazel; tarsi dark brown. Length 5·75. In winter black of head and throat is less pure. Female: feathers of head and black in centre with reddish-brown margins; throat whitish; white collar indistinct. Young: like female.

A common resident; in winter frequenting stubbles, etc., but in spring only found by streams, ponds, or in wet meadows and rush-covered pastures. Nest: in small bushes or rank herbage on sides of ditches and ponds; often quite on the ground; composed of fine dry rushes and grass-stalks, lined with finer grass and horse-hair. Eggs: 4 or 5; varying from purplish-grey to warm buff, with bold spots and streaks of deep purplish-brown; size ·80 by ·60.

Genus XLVIII. CALCARIUS, *Bechstein (1803).*

Bill and other characteristics much as in *Emberiza*, excepting that hind claw is longer and very little curved.

111. C. lapponicus (Linn.). Lapland Bunting.

Hab. Cimcumpolar regions, breeding chiefly within Arctic Circle. Migrating southward in winter.

Male: crown black, bordered by a broad white streak above each eye; sides of head, throat, and breast also black, margined by a white semi-collar which joins white eye-stripe; upon nape a broad band of light chestnut; remaining upper parts much as in Reed Bunting, but lighter and less chestnut-tinted; below white, sides boldly streaked with black; bill dull orange with a dusky tip; tarsi black. Length nearly 6·25. Female: head and cheeks brown streaked with black; feathers of mantle buffish-brown with dusky centres; chestnut band on nape scarcely noticeable, and white on sides of neck less distinct; throat white with a mottled blackish band on breast.

Of irregular but not uncommon occurrence in our south-eastern and eastern counties during autumn; has also been taken in most of the northern counties, twice in Scotland and once in Ireland, at Fastnet Rock (1887).

Genus XLIX. PLECTROPHENAX, *Stejneger (1882).*

Bill very short, strong, conical, lower mandible rather wider than upper. Other characteristics much as in *Emberiza*, excepting that the (curved) hind claw is considerably longer.

112. Plectrophenax nivalis (Linn.). SNOW-BUNTING.

Hab. Circumpolar regions, breeding northward of about lat. 58°. In winter, southward to about 35° N. lat.

Male: head, nape, and rump white, often showing blackish centres to some of the feathers; mantle and wings black, excepting greater coverts, bases of primaries and most of secondaries, which are white; three outer tail-feathers on each side white, rest black; below pure white; bill and tarsi black; iris hazel. Length 6·70. Female: black portions much duller, and white of head and nape mottled with blackish. In winter both sexes have upper feathers black in centre with broad margins of reddish-brown, while bill is orange with a black tip.

A common winter visitor to Scotland, the north and east of England, and north of Ireland; rare elsewhere. It has long been known to breed in the Shetlands, and has recently been found nesting in Sutherlandshire.

Family Sturnidæ.

GENUS L. STURNUS, *Linnæus (1766).*

Bill rather long, straight, base somewhat depressed, tip blunt. Wings long, broad, and pointed, 1st primary very small. Tail short. Feathers of head, nape, and breast narrow and pointed.

113. Sturnus vulgaris, Linn. STARLING.

Hab. Europe, north to lat. 70° in Norway; also W. Asia. In winter, partially migratory.

Male: plumage of body black, glossed with green, blue, and purple, most of feathers of upper parts and breast being tipped with buffish-yellow; wing and tail-feathers dusky, edged on outer margins with buffish-white; bill pale yellow; iris brown; tarsi reddish-brown. Length 8·25. Female: almost without glossy reflections and rather more spotted. Young: above sooty-brown; throat dull white; below mottled with dull brown and white; bill dusky.

Common and resident; in winter collecting in flocks of varying magnitudes. Nest: in holes in buildings, trees or cliffs. Eggs: 4 to 6; pale blue and very glossy; size 1·10 by ·85.

GENUS LI. PASTOR, *Temminck (1815).*

Bill moderate, compressed, upper mandible somewhat decurved and slightly notched. Feathers of crown lengthened and forming a crest.

114. Pastor roseus (Linn.). ROSE-COLOURED PASTOR.

Hab. Southern Europe and Western temperate Asia; casually to all parts of Europe.

Male: head, crest, nape, throat and upper breast black, glossed with purple; back, scapulars, and under parts white suffused with pale rose-colour; wings and tail glossy black; bill dusky at base, yellowish towards tip; tarsi dull orange. Length 8·50. Female: similar but duller.

A very irregular visitor in summer and autumn; has occurred in all parts of the British Isles.

Family Corvidæ.

GENUS LII. PYRRHOCORAX, *Vieillot (1816).*

Bill rather long, slender, ascending, and considerably curved, pointed at tip. Wings long and ample, 1st primary developed, 4th longest. Tail moderate. Claws large and curved.

115. Pyrrhocorax graculus (Linn.). Chough.

Hab. Mountains of Palæarctic Region, except extreme north, but very local.

Male: glossy black with purplish reflections; bill pale orange-red; iris hazel; tarsi light red; claws blackish. Length 15·00. Female: slightly smaller.

Locally distributed over rocky coasts of west side of Great Britain; also many localities around coasts of Ireland and in the Channel Islands. Nest: in crevices or caves of sea-cliffs or in ruined castles; composed of fine twigs, grass, wool and hair. Eggs: 3 to 5; pale greyish-yellow, mottled with light brown and dark grey; size: 1·46 by 1·10.

Genus LIII. NUCIFRAGA, *Brisson (1760).*

Bill moderately long, stout, straight, with a blunt point; wings moderate, rounded; tail moderate, nearly even.

116. Nucifraga caryocatactes (Linn.). Nutcracker.

Hab. Mountains of Europe (north nearly to Arctic Circle in Norway) but very local; also N. Asia.

Has occurred in nearly every eastern and southern county of England and twice in Scotland.

Genus LIV. GARRULUS, *Brisson (1760).*

Bill moderate, stout, compressed, tolerably straight; feathers of crown forming a crest; tail long rounded.

117. Garrulus glandarius (Linn.). Jay.

Hab. Europe, north nearly to Arctic circle.

Male: crest-feathers greyish-white, spotted with black; throat whitish, bordered each side by a black moustache-like stripe; rump white; rest of body pale brown with a rufous tint; wing-coverts chequered with black, light blue and white; quills black; primaries margined exteriorly with white, and secondaries with white bases, one next body, however, being reddish-brown; tail blackish; bill dusky-brown; iris bluish-white; tarsi light brown. Length 3·00. Female: similar.

Rather common and generally distributed, excepting in extreme north of Scotland; in Ireland very local, but breeds in most of south-eastern counties. Nest: in tall holly or thorn bushes, or up to 30 feet in oaks and firs; cup-shaped; composed of twigs lined with fine roots. Eggs: 5 or 6; pale greyish-green, closely mottled with pale brown, often with a few superficial black hair-streaks; size 1·25 by ·90. Note: a harsh screaming *kraark, kark.*

GENUS LV. PICA, *Brisson (1760).*

Bill moderate, stout, compressed, culmen decurved towards tip, which is indistinctly notched; wings rather short, rounded; tail very long, graduated.

118. Pica rustica (Scopoli). MAGPIE.

Hab. Northern Palæarctic region; in Europe north to lat. 70°.

Male: scapulars, part of inner webs of primaries, and plumage of abdomen pure white; on rump a band of grey; rest of plumage black, glossed with green and purple; bill and tarsi black; iris hazel. Length about 16·00. Female: plumage less glossed.

Fairly common in woodlands. Nest: in tall thorn bushes or 20 to 30 feet up in trees; composed of sticks, lined with roots and grass, and with a rough dome of sticks above. Eggs: 6; pale greyish-green or greyish-yellow, closely speckled with greenish-brown; size 1·35 by 1·05.

GENUS LVI. CORVUS, *Linnæus (1766).*

Bill moderately long, stout, strong, compressed, upper mandible rather longer than lower, and decurved towards tip; wings long, ample; tail moderate, rounded.

119. Corvus monedula, Linn. JACKDAW.

Hab. Eastern Palæarctic region; scarce in extreme south of Europe; occurs in N.W. Africa.

Male: nape and sides of neck light ash-grey; remaining plumage glossy sooty-black; bill and tarsi black; iris white. Length 13·00. Female: nape darker.

A common resident; generally distributed. Breeds in ruins, unused chimneys, hollow trees, or sea-cliffs; nest being a pile of sticks with a cup-shaped lining of straw, wool and feathers. Eggs: 5 or 6; greenish-white with distinct markings of dark olive-brown and lilac grey, rather sparingly distributed; size 1·45 by 1.05. Flight rather irregular. Note: sharp and querulous, uttered on the wing.

120. Corvus corone, Linn. CARRION-CROW.

Hab. Europe, excepting extreme north.

Male: whole plumage black, glossed with green and purple; bill black, clothed at base with reversed bristly feathers; tarsi black; iris grey. Length 18·00. Female: plumage less glossy.

Moderately common throughout Britain, excepting north of Scotland, but of very casual occurrence in Ireland. Abundant in Middlesex, placing its nest in all kinds of trees at a height of between 20 and 40 feet; nests are always found singly, unlike the Rook's, from which it may be distinguished by lining consisting of rope ends, bark-strips, wool, fur, hair, etc., often matted with mud; where trees are scarce it is placed on rock-ledges. Eggs 3 to 5; dull pale green, boldly spotted or blotched with dark brown; size 1·68 by 1·20. Quite omnivorous.

121. Corvus cornix, Linn. HOODED CROW.

Hab. Northern, Central and South-Eastern Europe; also N.E. Africa and W. Asia.

That this is very closely allied to *C. corone* is made evident by the fact that the two constantly inter-breed while the hybrids even appear to prove fertile. The pure *C. cornix*, however, has the back, breast and abdomen light ash-grey with fine dusky shaft-streaks to most of the feathers, the remaining plumage being like that of *C. corone.*

Breeds throughout Scotland, Ireland, and the Isle of Man, but very rarely in England, even in the north, although small flocks from North Europe occur everywhere during winter. Nidification is similar to that of last species but eggs are slightly larger; size 1·72 by 1·25. Like the Carrion-Crow it is a bold marauder.

122. Corvus frugilegus, Linn. ROOK.

Hab. Europe (excepting extreme south); also whole of Western Asia.

Male: whole plumage glossy bluish-black; skin of forehead and upper throat destitute of feathers and ashy-white in colour; bill and tarsi, black. Length about 17·50. Female similar. Young: skin round base of bill is at first clothed with feathers.

Common and resident everywhere, breeding usually in the familiar rookeries in tree-tops; occasionally a quite isolated nest is found, while where trees are lacking it breeds in bushes or on the ground. Nest: lined with a quantity of grass and fibrous roots. Eggs: 4 or 5; similar to those of *C. corone* but usually with smaller and closer markings; also not so large; size 1·60 by 1·18.

123. Corvus corax, Linn. RAVEN.

Hab. Northern Palæarctic region.

Male: whole plumage black, glossed with purple, especially on the long pointed feathers of the throat; bill and tarsi black; iris grey. Length 23·00. Female: duller and a trifle smaller.

Still breeds in almost every part of British Isles, excepting eastern and central portions of England and some parts of Ireland; most abundant in Scotland. Nest: in cliffs or trees; consists of a quantity of sticks surmounted by a cup of wool, hair, etc.; when in cliffs it is often used for many years and added to annually. Eggs: 3 to 5; similar to those of *C. corone* but larger; size 1·92 by 1·35. A noted robber and absolutely omnivorous.

Family Alaudidæ.

GENUS LVII. ALAUDA, *Linnæus (1766).*

Bill of moderate length, somewhat slender, nearly straight; upper mandible arched and slightly decurved at tip, but not notched. Wings long, tertiaries much lengthened. Hind claw long and almost straight.

124. Alauda arvensis, Linn. SKY-LARK.

Hab. Whole Palæarctic region; in Europe north, locally, to lat. 70°. Partially migratory in winter.

Male: feathers above dark brown in centre with broad buff margins; above eye a short whitish streak; feathers of crown forming a short semi-erectile crest; outer tail-feather each side mainly white, next one margined exteriorly with white; below warm buffish-white, boldly spotted on throat and sides of breast with dark brown; bill dark brown, yellowish at lower base; iris hazel; tarsi pale brown. Length about 7·00. Female: similar but a trifle smaller, and with a very slight crest. Young have a pronounced buff tint.

Common everywhere; usually frequenting cultivated land; in winter gregarious. Nest: on ground, among grass, clover, wheat, etc.; composed of grass, lined sometimes with horsehair. Eggs: 3 to 5; greyish-white, closely spotted and mottled with warm olive brown; size ·95 by ·65. The glad song, uttered while soaring upward is well known; it may be heard at nearly all seasons.

125. Alauda arborea, Linn. WOOD-LARK.

Hab. Europe, south of lat. 60° N., but rather local. In winter partially migratory.

May be distinguished from Sky-Lark, even in the field, by smaller size and much shorter tail and wings, former also showing less white on outer feathers, but having nearly all tipped with white; under parts are yellower, first or "bastard" primary larger, bill noticeably weaker and stripe over eye broader and longer, running from in front of eye to nape. Length 6·00; female slightly smaller.

Locally distributed over most parts of England and Wales; breeds casually in South Scotland, but in Ireland is even more rare than formerly, except in winter, and only breeds exceptionally in the south. Nest: on ground, in a grass-tuft or at foot of a bush; more substantial than Sky-Lark's. Eggs: yellowish or greenish-white, rather sparingly marked with small spots of reddish-brown and lilac-grey; size ·85 by ·65. Sweet and liquid, but simple song is uttered both on the wing and while perched.

126. Alauda cristata, Linn. CRESTED LARK.

Hab. Europe, north in east to lat. 60°; also N.W. Africa and Western Asia.

Male: greyer and duller above than *A. arvensis*, with a very distinct and pointed crest, and a broad eyestripe; wings and tail shorter, and the latter *without white* in it. Length nearly 7·00. Female slightly smaller. It is more nearly allied to *A. arborea* than to *A. arvensis*, and, like former, has the bastard quill much larger than in latter.

A rare autumn visitor. Five have been taken in Cornwall (one of them in June), and two in Sussex; there are other unauthenticated records.

GENUS LVIII. CALANDRELLA, *Kaup (1829).*

127. C. brachydactyla (Leisler). SHORT-TOED LARK.

Hab. South Europe, N.W. Africa, and Central Asia.

A rare visitor on migration. One has been taken near Shrewsbury, one near Cambridge, one in Scilly Isles, one at least near Southampton, two or three near Brighton, and one in Ireland.

GENUS LIX. MELANOCORYPHA, *Boie (1828).*

128. M. sibirica (Gmel.). WHITE-WINGED LARK.

Hab. North-East Europe and Western Asia.

A single example was taken near Brighton in November 1869.

GENUS LX. OTOCORYS, *Bonaparte (1839).*

Bill much as in *Alauda* but slightly shorter. Male with a pointed tuft of feathers capable of erection on each side of the crown. Tail moderately long. Claws rather long and but little curved; hind one nearly as long as in *Alauda*.

129. Otocorys alpestris (Linn.). SHORE-LARK.

Hab. Northern Palæarctic region (within the Arctic circle); also Greenland and Eastern Boreal America. In winter migrating southward.

Male: forehead, sides of head, and upper throat yellowish-white, enclosing a black patch on lores and ear-coverts; fore-part of crown and the "horns" blackish; rest of upper parts light-brown, with a slight red tinge; greater coverts tipped with white; two outer tail-feathers margined with dull white; upper breast with a deep crescent of black; lower breast slightly streaked; remaining under parts whitish, with some streaks on sides; bill dusky; tarsi black. Length 6·75. Female: slightly smaller, duller, and horns are absent.

An almost annual visitor in spring or autumn, sometimes in flocks, to east side of Britain and also to the south coast; as yet unknown in Ireland.

ORDER PICARIÆ.

Family Cypselidæ.

GENUS LXI. CYPSELUS, *Illiger (1811).*

Bill very short, base wide, depressed, tip compressed; upper mandible noticeably decurved, lower slightly so; gape very wide. Wings very long, narrow, pointed; tail moderate, forked. Tarsi short, feathered to toes, latter four in number, directed forward, but capable of grasping in opposition; claws strong, much curved.

130. Cypselus apus (Linn.). SWIFT.

Hab. Palæarctic region, north to lat. 70° in Norway. In winter southward to Africa and India.

Adult: chin greyish-white; rest of plumage sooty-brown, with slight reflections; bill, toes and claws black. Length from base of bill to tip of tail 6·50. Young: browner and with dull white of chin more extended.

Common from beginning of May to early in August, making its slight nest in holes in thatched roofs, cliffs, church towers, etc., and laying two dull white eggs of an elongated oval shape; size ·98 by ·62. Only note is a loud screech, uttered as the birds sweep overhead—higher up than either Swallow or Martin.

131. Cypselus melba (Linn.). ALPINE SWIFT.

Hab. Southern Palæarctic region, eastward to India. Migrates southward in winter.

Adult: throat and belly white; remaining plumage, including a band across breast, deep brown with slight reflections on wings; lores black; bill black; toes pale brown. Length from base of bill to tip of tail fully 7·50.

Has occurred about twenty times in England and Wales, but is not recorded from Scotland; in Ireland three examples have been taken.

GENUS LXII. ACANTHYLLIS, *Boie (1826).*

132. Acanthyllis caudacuta (Lath.). NEEDLE-TAILED SWIFT

Hab. Eastern Siberia, China and the Himalayas. In winter reaching Eastern Australia.

An example was shot at Great Horkesley, near Colchester, Essex, in 1846, and another near Ringwood, Hants, in 1879.

Family Caprimulgidæ.

GENUS LXIII. CAPRIMULGUS, *Linnæus (1766).*

Bill very short, base wide, tip compressed, tip of upper mandible decurved; gape very wide, beset above with large bristles. Wings long. Tail rather long, rounded. Tarsus short, feathered nearly to toes anteriorly; three toes in front, one behind; claw of middle toe larger than others and with its inner edge serrated.

133. Caprimulgus europæus, Linn. NIGHTJAR.

Hab. Europe, north to S. Scandinavia; also Western Asia. In winter south to Africa and India.

Male: chiefly light grey, diversely barred and vermiculated with blackish-brown and rufous; wings barred with reddish-white and with an oval white spot on inner webs of three outer primaries at about middle; two outer tail-feathers on each side also have a large spot of white at tip; throat irregularly banded with white markings; bill brownish-black; feet brownish-orange. Length fully 10·00. Female: lacks white spots on wings and tail.

Common from latter part of May to end of September. Frequents chiefly beds of bracken, furze-covered commons or coniferous woods, the 2 oval eggs being laid in a slight depression in the ground; they are creamy-white, blotched or marbled with deep brown and lilac-grey; size 1·25 by ·82. The vibrating and sustained "jar" or "churr" is well-known; it is probably never uttered on the wing. Feeds on winged insects, such as the "cockchafer"; is always observed to perch lengthwise on the larger branches of trees, never across them, but seems preferably to rest on the ground.

134. Caprimulgus ruficollis, Temm. REDNECKED NIGHTJAR.

Hab. S.W. Europe and N.W. Africa.

A single occurrence near Newcastle in 1856 was recorded by the late John Hancock.

135. Caprimulgus ægyptius, Licht. EGYPTIAN NIGHTJAR.

Hab. N.E. Africa and S.W. Asia.

An example in the collection of Mr. Whitaker was shot at Rainworth, Notts., June 23rd, 1883.

Family Picidæ.

Sub-Family Picinæ.

GENUS LXIV. DENDROCOPUS, *Koch (1816).*

Distinguished from *Gecinus* largely by black and white colours and small size.

136. Dendrocopus major (Linn.). GREAT SPOTTED WOODPECKER.

Hab. Europe, north to Arctic Circle; also temperate Asia.

Male: forehead buffish-white; crown black; sides of head and the throat white, a stripe of black dividing the two; occiput crimson; rest of upper parts black with a white patch on each side of neck; scapulars pure white; primaries and secondaries irregularly barred exteriorly with white; tail-feathers, except two middle ones, tipped with white, the outer ones showing very much white; under parts dull white, excepting lower tail-coverts, which are crimson; bill greyish-black; iris crimson; tarsi olive-grey. Length about 9·25. Female: crown buffish-white; occiput black. Young: like female, but crown is suffused with red.

Resident in England and Wales, breeding sparingly in most districts in the older woodlands. It appears also to have bred in Scotland, but is chiefly known as an uncertain winter visitor; also visits Ireland at same season. Nesting hole is usually excavated by the birds themselves in tree-trunks at varying heights. Eggs: 5 to 7; shell glossy; creamy-white; size 1·00 by ·75. Food: chiefly insects and small larvæ, with berries or nuts in winter.

137. Dendrocopus minor (Linn.). Lesser Spotted Woodpecker.

Hab. Northern Palæarctic region; in Europe north to North Cape.

Instantly distinguished from *D. major* by its much smaller size, the adult measuring barely 5·50. The male has a good deal of white on the sides of neck and the back; the nape black and crown light crimson. Female has crown buffish-white.

A common resident in south of England, but very rare in the northern counties, and seldom recorded from Scotland; six or seven examples have been obtained in Ireland. Chiefly frequents woods, but I have found the nest in small decayed stumps about the meadows in Middlesex. Breeding hole is always excavated by the birds; in height its situation ranges from 2 to 30 feet. Eggs: 5 to 7; shell glossy; creamy white; size ·75 by ·55. Call-note: a very loud and shrill *plee, plee, plee, plee, plee*, uttered as the bird clings to a tree.

Genus LXV. GECINUS, *Boie (1831)*.

Bill moderately long, nearly straight, stout at base, tapering towards tip, which is obtusely pointed, upper mandible being very slightly longer than lower and a little arched from near the base; tongue cylindrical, long, allowing of considerable projection, the tip barbed. Tail pointed of 12 feathers, all but short outer pair having shafts very stiff and with projecting tips. Toes two in front and two behind; claws large, much curved.

138. Gecinus viridis (Linn.). GREEN WOODPECKER.

Hab. Western Palæarctic region, north to South Scandinavia, south to Mediterranean and Pyrenees.

Male: top of head and a moustache-like streak from gape crimson, the sides of head being black; mantle dull light green; rump yellow; primaries dusky, barred exteriorly with buffish white; below light greyish-green; bill blackish; iris ash-white; feet greyish. Length about 12·00. Female: red of cap less extensive and that of moustache absent.

Common and resident in most parts of England and Wales, but has not occurred half-a-dozen times in Scotland, while it had only been taken twice in Ireland previously to October, 1889, when an extensive immigration occurred. Nesting burrow is excavated in trunks or larger limbs of trees at variable heights; entrance about 2·50 in diameter and interior cavity rather deep. Eggs: 5 to 7; shell glossy; creamy white; size 1·30 by ·90. Food: chiefly insects taken upon tree trunks, but in both summer and winter it may be seen at work upon the ant-hills on the ground.

Sub-family, Iynginæ.

GENUS LXVI. IŸNX, *Linnæus (1766).*

Bill moderate, straight, somewhat conical, terminal half diminishing rapidly, tip sharply pointed; tongue as in *Gecinus*, but the tip without barbs. Tail of ten feathers, tips webbed beyond shaft, and not stiff. Feet as in *Gecinus*. Feathers of crown forming a semi-erectile crest.

139. Iÿnx torquilla, Linn. WRYNECK.

Hab. Palæarctic region, excepting extreme north. In winter southward to Africa, India and China.

Male: above pale ash, mottled and vermiculated with darker grey and brown; feathers of crown barred with brownish-black, and nape, centre of mantle, and scapulars streaked with same; tail crossed by several irregular bands of brownish-black; throat buff with fine transverse bars of blackish-brown; under parts dull white, more or less marked with dusky spots and bars; bill and tarsi brownish. Length nearly 7·00. Female: similar.

Common throughout southern and midland counties, but scarce in northern counties and in Wales; to Scotland a very irregular visitor; in Ireland has occurred twice. About April 1st in the southern counties it makes known its arrival by its loud vociferous call-note. The 7 or 8 eggs are laid in holes in trees, but it never excavates its own nesting-hole; they are larger and more oval than those of *D. minor*, shell being much less glossy and of a dead white; size ·83 by ·62. Feeds largely on ants.

Family Alcedinidæ.

GENUS LXVII. ALCEDO, *Linnæus (1766).*

Bill long, almost straight, stout and wide at base, diminishing to an acute point. Wings rather short, rounded; tail of 12 feathers, very short. Legs bare above tarsal joint; tarsus rather short; toes rather small, three in front, one behind, the fore toes united at base by a membrane.

140. Alcedo ispida, Linn. KINGFISHER.

Hab. Europe, south of lat. 60°; also N.W. Africa.

Adult: crown, nape, wing-coverts, and a moustache-like stripe below the chestnut side of head deep greenish-blue, barred with azure-blue; back and upper tail-coverts brilliant azure-blue; wing-quills and tail dark greenish-blue, secondaries having rufous outer margins; chin and throat dull white; under parts light chestnut; bill orange at base, black towards tip; iris hazel; tarsi dull pale red; claws blackish. Length about 7·00.

Sparingly distributed and resident, excepting in the north of Scotland; in Ireland it is scarce and local, although breeding in nearly every county. Nesting burrow is often excavated by the bird in the bank of a stream or river, but sometimes a ready-made hole is used. From 6 to 8 eggs are laid on bare soil or often on fish-bones cast up by the bird; rather globular in shape with a glossy white shell; size ·90 by ·75.

GENUS LXVIII. CERYLE, *Boie (1828).*

141. Ceryle alcyon (Linn.). BELTED KINGFISHER.

Hab. North America. Partially migratory in winter. Two examples preserved in Dublin were shot in Ireland in 1845, one in Co. Meath, the other in Co. Wicklow. Although the B.O.U. Committee admitted it to the British list on the strength of these occurrences, several authorities do not regard it as a possible visitor. Those, however, who are acquainted with the powerful flight and *migratory* habits of this somewhat aberrant type of Kingfisher must allow it to be a *possible* although certainly not a *probable* visitor.

Family Coraciidæ.

GENUS LXIX. CORACIAS, *Linnæus (1766).*

142. Coracias garrula, Linn. ROLLER.

Hab. Western Palæarctic region, in continental Europe breeding northward to Southern Scandinavia. Migrating in winter to South Africa.

A rare and irregular migrational visitor to the British Isles. A considerable number of occurrences are on record.

Family Meropidæ.

GENUS LXX. MEROPS, *Linnæus (1766).*

143. Merops apiaster, Linn. BEE-EATER.

Hab. South Europe, North Africa and S.W. Asia. Migrating to South Africa in winter.

An irregular visitor, on migration, to southern half of England; has been recorded from Scotland on four occasions, while eight examples have been shot in Ireland, two of them from small flocks.

Family Upupidæ.

GENUS LXXI. UPUPA, *Linnæus (1766).*

Bill long, slender, slightly decurved, laterally compressed and tapering to a slender point. Feathers of head long, forming an erectile crest. Tail even, moderately long. Tibia feathered; tarsus moderate, scutellated; three toes in front, one behind.

144. Upupa epops, Linn. HOOPOE.

Hab Palæarctic region, excepting extreme north. In winter southward nearly to Equator.

Male: general tint light reddish-buff, the graceful crest being tipped with black, the back banded with black and white, and the wings and tail black barred with white; bill black, paler at base. Length 9·75. Female similar.

An almost annual visitor in small numbers to south of England in spring; sometimes in autumn; has bred at long intervals in every county from Kent to Devon. Also occurs in small numbers nearly every year in south of Ireland, but is a rare straggler to Scotland.

Family Cuculidæ.

Genus LXXII. CUCULUS, *Linnæus (1766).*

Bill somewhat short, moderately wide at base, compressed towards point, upper mandible decurved at tip and with an apparent notch, lower straight. Wings and tail rather long, latter a little graduated. Tarsus short, feathered on upper part; two toes in front and two behind.

145. Cuculus canorus, Linn. Cuckoo.

Hab. Whole Palæarctic region. In winter migrating to South Africa and India.

Male: above rather dark grey, throat paler; lower breast and under parts white, with transverse blackish bars; tail tipped and slightly spotted with white; bill dusky, yellowish at base; iris, feet and claws yellow. Length 12·50. Female similar. Young have the grey tint replaced by dark brown with darker markings; iris brown; feet pale yellow. The variety of a pale rufous tint with darker bars (formerly considered a distinct species) occurs in both sexes, but chiefly or wholly as their second season's plumage.

Common from Mid-April to August or September throughout whole of British Isles. Male announces his arrival by his well-known powerful and musical note which he delivers frequently on the wing as well as when perched; when in the act of settling on a tree he also often utters a low harsh rattling note. The female lays her eggs in the nests of other birds, and undoubtedly always deposits them in the selected nest by means of her bill, the small size of the egg rendering this possible; several eggs are placed in this way in different nests. Eggs vary from greenish-white to pale rufous-grey, spotted and mottled closely or otherwise with olive-brown or rufous-brown; sometimes of a pale unspotted blue; size ·90 by ·75. Owing to amount of food necessary for its own development the young Cuckoo invariably ejects its foster-brethren from the nest; it is said that this occurs when it is 9 or 10 days old, but in my experience it takes place about the third day, when the Cuckoo is both blind and naked.

GENUS LXXIII. COCCYSTES, *Gloger (1834).*

146. Coccystes glandarius (Linn.). GREAT SPOTTED CUCKOO.

Hab. S.W. Europe and North Africa.

Has occurred once in Co. Galway, Ireland, March, 1842, and once on the Tyne, August, 1870; the first example is in Trinity College Museum, Dublin, and the second in the Newcastle Museum.

GENUS LXXIV. COCCYZUS, *Vieillot (1816).*

147. Coccyzus americanus (Linn.). YELLOW-BILLED CUCKOO.

Hab. America, from Canada southward to Brazil. Migrating from northern portions in winter.

Several authorities have doubted whether this is a genuine visitor, but as its occurrences have all been at the season of autumn migration, and have also been in the quarters where a genuine visitor is most likely to occur, it seems desirable to retain it here. It has occurred twice in Ireland (in Cork and Wicklow), once in in Cornwall, once in Pembrokeshire, and once near Aberystwith.

148. Coccyzus erythrophthalmus (Wils.). BLACK-BILLED CUCKOO.

Hab. America, from Canada to Brazil. Migrating from northern portions in winter.

A single example was shot in Co. Antrim, Ireland, September 25th, 1871. The remarks made above apply also to this species.

ORDER STRIGES.

Family Strigidæ.

In the owls the outer toe is reversible, two toes being invariably directed forward and two behind when perching. I know this to be the case in *Strix*, *Bubo*, *Syrnium*, *Athene*, and *Nyctea*, and it is doubtlessly so in all other genera.

GENUS LXXV. STRIX, *Linnæus (1766).*

Bill moderate, basal portion straight, upper mandible much decurved at tip, but with cutting edges nearly straight; lower mandible notched; facial disc complete; opening of the ear very large. Wings long; tail short, nearly even. Tarsus rather long, feathered; toes naked, save for a few hairy feathers; claws, long, curved.

149. Strix flammea, Linn. BARN-OWL.

Hab. Europe (excepting eastern Russia), north to lat. 60° in the west; also North Africa from Morocco to Egypt.

Male: above buffish-orange, with small longitudinal spots of brownish-black, and white and delicate grey pencillings; facial discs white, edged with reddish-brown; primaries and tail barred with greyish-brown;

under parts white; bill whitish; iris nearly black; feet clothed with white downy feathers; claws pale brown. Length 13·25. Female: length nearly 14·00. In a dark phase not uncommon in the east and south of England, particularly in autumn, the orange tint of upper parts is largely obscured by a smoke-grey hue and under parts are suffused with buffish-yellow, and marked with a number of the small dusky-grey spots sometimes noticed in normal examples.

Resident, and widely but sparingly distributed. Resorts chiefly to hollow trees, old barns, church towers and ruins. Eggs are usually found upon a quantity of the birds' ejected pellets; from 4 to 6 or more are laid at intervals, the first being much incubated before last are laid; shape oval, of a dull surface, and white; size 1·60 by 1·25. Young are fed almost entirely on mice.

Family, Asionidæ.

Genus LXXVI ASIO, *Brisson (1760).*

Bill with upper mandible decurved from its base; head furnished with an erectile tuft of feathers on either side; toes clothed with short downy feathers.

150. Asio otus (Linn.). Long-eared Owl.

Hab. Palæarctic region, except extreme north.

Male: above yellowish-buff (darker on wings and rump) mottled and pencilled with brown and grey, and longitudinally streaked with brownish-black; tail rufous with dusky markings and bars; long ear-tufts black in centre, with pale edges; facial discs dusky round eyes, rest pale buff with a dusky rim; below greyish mottled with pale brown, with longitudinal streaks and fine transverse bars of blackish-brown; bill and claws nearly black; feet clothed with short yellowish-brown feathers; iris yellowish-orange. Length 14·00. Female: larger.

Resident and generally distributed; fairly common in woodlands. Eggs are laid in old nests of other birds, commonly a Ringdove's or Crow's, or Squirrel's dreys; usually 4 or 5 in number; white; in shape oval with a smooth dull shell; size 1·65 by 1·30.

151. A. brachyotus (Forster). SHORT-EARED OWL.

Hab. Almost cosmopolitan; in Europe breeding up to the North Cape, but locally distributed.

Distinguished instantly from *A. otus* by its short ear-tufts, these measuring only about ·70 instead of 1·30 as in the latter species; they are not normally kept erected. The facial discs and upper parts in general are darker than in *A. otus*, and under parts are buffish-white, streaked longitudinally, but not barred, with dusky brown. Length: male 14·00; female 14·75.

A regular winter visitor in varying numbers, but in Great Britain found breeding locally on the moors and fens from East Anglia northward and throughout Scotland. Eggs: 4 or 5; like those of *A. otus*, but of a slightly narrower oval shape; size 1·60 by 1·25.

GENUS LXXVII. SYRNIUM, *Savigny (1810).*

Head large, without ear-tufts; wings rather short, 2nd, 3rd and 4th primaries graduated, latter longest; tail moderately long; toes feathered to claws.

152. Syrnium aluco (Linn.). TAWNY OWL.

Hab. Western Palæarctic region.

Male: above tawny-brown mottled with grey, and with darker, almost black, markings; wing-coverts marked on outer webs with conspicuous white spots; tail barred (except two central feathers) with dark brown; facial disc greyish-white with a narrow dark brown rim; under parts greyish-white, mottled with buffish-brown and longitudinally streaked with dark brown; bill and claws whitish; feet clothed with greyish-white feathers; iris dark brown. Length 15·00. A grey phase sometimes occurs in which the tawny hue is largely replaced by smoke-grey. Female: length 16·00.

Resident and tolerably common in Great Britain, but has not yet been identified in Ireland. For breeding purposes hollow trees are most commonly resorted to, but it has been known to make use of old nests of Crows, etc., and holes in ruins. Eggs: often 4; nearly globular, of a smooth dull surface and white; size 1·80 by 1·55. The common "hooting" owl; of distinctly nocturnal habits.

GENUS LXXVIII. NYCTEA, *Stephens (1826).*

Facial discs very incomplete; ear-tufts almost absent; openings of ears not large; wings moderate, 3rd primary longest; tail moderate, ample, rounded; tarsi and toes clothed with large feathers.

153. Nyctea nyctea (Linn.). SNOWY OWL.

Hab. Circumpolar regions; north of lat. 60° in Europe.

Male: snowy white, barred and spotted with blackish-brown; bill and claws black; iris bright yellow. Length 23·00. Female: larger and decidedly more heavily barred. Length 26·00.

Occurs almost every winter in North Scotland, Orkneys and Shetlands; also at intervals down the east side of England, but only once in the south. In Ireland eight examples have been shot, four of them in Co. Mayo.

Genus LXXIX. SURNIA, *Duméril (1806).*

154. S. ulula (Linn.). European Hawk-Owl.

Hab. Northern Europe. Migrating to Central Europe in winter.

A single example has been taken near Amesbury, Wilts (R. B. Sharpe, P.Z.S. 1876, p. 334).

154a. S. ulula caparoch (Linn.). American Hawk-Owl.

Hab. Northern Nearctic region. In winter southward to Northern United States.

Most of the Hawk Owls taken in Britain appear to belong to this form. It has occurred three or four times in Scotland, once in Cornwall (1830), and once in Somerset (1847).

Genus LXXX. NYCTALA, *Brehm (1828).*

Head rather large; without ear-tufts; openings of ears large; facial discs tolerably apparent; wings fairly long; tail short. Tarsus and toes clothed with tolerably large feathers.

155. Nyctala tengmalmi (Gmel.). Tengmalm's Owl.

Hab. Northern Palæarctic region.

Male: above umber-brown, with large oblong white spots on the wing-coverts and mantle, and smaller drop-shaped white marks on the head; facial discs greyish-white with a dark rim; below whitish tinged with buff, irregularly banded on breast and striped on flanks with dark brown; bill horn-white; toes clothed with mottled buff and white feathers. Length 8·50. Female: a little larger.

About sixteen examples have been taken in various parts of England, and two in Scotland.

GENUS LXXXI. SCOPS, *Savigny (1809).*

Head rather small, with fairly conspicuous ear-tufts; facial discs very incomplete; openings of ears small. Wings long. Tarsi covered with short feathers; toes bare. Size small.

156. Scops scops (Linn.). SCOPS OWL.

Hab. Southern Europe, North Africa, and S.W. Asia.

A rare straggler in spring or autumn; it has occurred at long intervals in many parts of England, once or twice in Wales, once in Scotland, and four or five times in Ireland.

GENUS LXXXII. BUBO, *Duméril (1806).*

157. Bubo bubo (Linn.). EAGLE OWL.

Hab. Europe and N.W. Africa.

Probably at one time breeding in the Orkneys which, with the Shetlands, it still visits at considerable intervals; one or two have been taken on the east coast of Scotland and in England, but none of them recently.

GENUS LXXXIII. CARINE, *Kaup (1829).*

Head large, round, without ear-tufts; openings of ears moderate; facial discs incomplete. Tarsus moderately long, clothed with short downy feathers; toes covered sparingly with very short hairy feathers.

158. Carine noctua (Scop.). LITTLE OWL.

Hab. Southern half of Europe.

Has been taken at intervals in nearly all parts of England; and has also bred in several counties, but it is known that numbers have been liberated here and the nests found can be attributed to these, as in its wild state the bird is probably only a rare struggler to us.

From a similarity of size this Owl and *N. tengmalmi* are frequently confused with one another, but the present species may be distinguished by its toes being merely sparingly covered with short whitish hair-like feathers.

ORDER ACCIPITRES.

Family Vulturidæ.

GENUS LXXXIV. GYPS, *Savigny (1810).*

159. **Gyps fulvus** (Gmel.). GRIFFON VULTURE.

Hab. Southern Europe, North Africa and S.W. Asia.

An example in the Trinity College Museum, Dublin, was captured at Cork in 1843.

GENUS LXXXV. NEOPHRON, *Savigny (1810).*

160. **Neophron percnopterus** (Linn.). EGYPTIAN VULTURE.

Hab. Southern Europe, Africa, and S.W. Asia.

A very rare straggler. One of a pair was shot at Kilve, Somersetshire, in October, 1825, and another at Peldon, Essex, 28th September, 1868.

Family Falconidæ.

GENUS LXXXVI. CIRCUS, *Lacépède (1800).*

Bill moderate, compressed, upper mandible decurved from base and indistinctly toothed; cere very apparent; head exhibiting partial facial discs. Wings very long, usually reaching to end of tail, which is also long. Tarsi long, slender, bare; claws moderately curved.

161. **Circus æruginosus** (Linn.). MARSH-HARRIER.

Hab. Palæarctic region. Migrating from northern regions in winter.

I 2

Male: head and nape yellowish-white slightly streaked with deep brown; feathers of back dark brown with narrow lighter margins; primary quills blackish-slate; remainder of wings and the tail pale ash-grey; chin and throat buffish-white; under parts rufous-buff, the whole streaked with dark-brown; bill blackish; irides, cere and feet yellow; claw black. Length 20·00. Immature males are chiefly umber-brown with a brownish-white pate; iris hazel. Female: wings and tail brown. Length 22·00.

Now a rare British bird, although formerly breeding in many parts of England; at the present time Norfolk seems to be the only county in which a few still nest. To Scotland it is a rare straggler only, while it has been exterminated in most counties of Ireland, although Mr. Ussher reports it as still breeding in Queen's Co., Galway, and probably King's Co. and Westmeath.

162. **Circus cyaneus (Linn.).** HEN-HARRIER.

Hab. Northern Palæarctic region, north to within Arctic circle. Partially migratory in winter.

Male: whole plumage pale bluish-grey, excepting rump and abdomen, which are nearly white, and primaries, which are dusky-slate; bill and claws blackish; irides, cere, and tarsi yellow. Length 18·50. Female: brown, darkest above, streaked with white on nape and on edges of the slight facial discs, and much mottled with white on rump; tail barred with blackish-brown, and under parts streaked with same; cere greenish-yellow; iris reddish-brown. Length 20·50. Young: like female; males in second year are much like old birds, excepting that tail shows indistinct dusky bars, and nape and sides of abdomen are somewhat streaked with brown.

A decreasing species, but still to be found breeding in a few localities in the west, south-west, and north of England, also in Wales and more commonly in the Orkneys, Hebrides, and some parts of the Highlands; in Ireland it breeds sparingly in Kerry, Galway, and possibly other counties. Nest: commonly placed in tall heather but often on more open ground; consists of heather twigs and dry grass. Eggs: 4 or 5; bluish-white, usually unmarked, but now and again with faint orange-brown marks; size 1·78 by 1·42.

163. Circus cineraceus (Montagu). MONTAGU'S HARRIER.

Hab. Europe, except extreme north; also Western Asia. Migrates southward in winter.

Male: smaller, more slender, and of a noticeably darker grey than the male of *C. cyaneus*; the tail is also distinctly barred in the adult bird, the secondary wing-quills exhibit a dusky band, and the under parts are white streaked with reddish-brown; the wings are very long, reaching *beyond* the tip of the tail. Length 17·50. Female: much like female of *C. cyaneus*; often more rufous below and darker generally. Length 19·00. Young: like female.

A summer visitor to England, but much less common than formerly. It has bred irregularly in the south-west from Hants to Devon, also in Norfolk and parts of Wales; to the north and to Scotland it is a very rare straggler; in Ireland five examples have been obtained in Wicklow, Wexford, and Co. Dublin. Nest: on the ground among heather, gorse, rushes, etc. Eggs: 4 or 5; like those of *C. cyaneus*, but smaller; size 1·68 by 1·35.

Genus LXXXVII. BUTEO, *Lacépède (1800).*

Bill moderate, upper mandible decurved from the base, but with the tooth in cutting edges almost obsolete ; tarsus short and stout, naked.

164. Buteo buteo (Linn.). Buzzard.

Hab. Europe.

Adults vary extremely in plumage. Both sexes are sometimes found of a nearly uniform blackish-brown above and below, this being, according to some authorities, the plumage of very old birds. Adult birds may, however, be met with (especially on the Continent) which have more white than brown in their plumage, the male indeed, being pure white, with bold brown blotches on upper parts, and the normal dark bars on tail. I have seen an example in this state of plumage, which was shot in Notts. Young birds, in first year's plumage, are mottled with brown and yellowish-buff above ; below buffish-white, usually mottled with brown, but sometimes unmarked, tail being rufous-grey, with darker bars. Length of adults: male 20·50 ; female 22·50 ; bill and claws bluish-black ; iris, cere and tarsi yellow.

Formerly bred throughout our islands, but now confined to various localities on the west side of Great Britain from Wales to the Inner Hebrides. In Ireland very scarce, but may still breed in Donegal and Londonderry. Nest is built in cliffs in mountainous districts, but also commonly in trees. Eggs: 3 or 4 ; dull white, slightly blotched or streaked with rust-brown ; size 2·20 by 1·75. By no means a courageous bird ; feeding principally on mice, frogs, large insects, and small birds. Note: a plaintive squeal.

Genus LXXXVIII. ARCHIBUTEO, *Brehm (1828).*

Bill and appearance approximate more to *Aquila* than to *Buteo*. Tarsus feathered to origin of toes.

165. A. lagopus (Gmel.) Rough-legged Buzzard.

Hab. Northern Europe and Western Siberia.

May be easily distinguished by the feathered tarsus; it is larger than *B. vulgaris*, and the adult has a conspicuous blackish-brown patch covering centre of abdomen, the rest of under parts being chiefly brownish-white; tail also shows only two or three dark bands on its terminal half, the base being nearly white. Length: male 22·50; female 24·50; iris hazel in adults but yellow in immature examples.

An irregular, but probably annual, winter visitor to the east side of Great Britain but rarer in the west; in Ireland seven or eight examples in all have been taken.

Genus LXXXIX. AQUILA, *Brisson (1760).*

Bill moderate, stout, upper mandible nearly straight at base, terminal half much decurved and hooked, cutting edges slightly waved but not toothed. Tarsi clothed with feathers down to origin of toes, which are reticulated above, excepting last joint which is covered with three broad scales; claws large, hooked.

166. Aquila maculata (Gmel.). Larger Spotted-Eagle.

Hab. Central and S.E. Europe; also S.W. and Central Asia. In winter reaching Abyssinia.

Two examples have been taken in Ireland (1845), two in Cornwall (1860-61), one in Lancashire (1875), and one in Northumberland (1885). It should be said that a smaller form exists under the name of *A. nævia* (more correctly *A. pomarina* of Dresser), inhabiting Germany, Poland,

and Central Russia, and as many of our ornithologists refuse to recognise anything but *species*, the name of this smaller race has been often applied to the British examples, but it is the larger form *only* which has occurred in Britain, as vouched for by Mr. Gurney (Ibis, 1877, p. 332) and Mr. Dresser. (Zoologist, 1885, p. 230).

167. Aquila chrysaëtus (Linn.). GOLDEN EAGLE.

Hab. Palæarctic and Nearctic regions.

Male: whole plumage deep brown, inclining to tawny red on the head and nape, and to reddish-brown on the belly and thighs; bill dusky; iris hazel; cere and toes yellow; claws black. Length 31·00 to 32·00. Female: length 35·00 to 36·00. Young: base of tail white, some white mottlings also often appearing on breast; white on tail is probably not lost until third or fourth year.

Now only known to breed in the Highlands, the Hebrides, and more rarely in one or two localities in the west and north of Ireland. In its Scotch haunts it is likely to continue, since it is being protected by many owners of deer-forests. The eyrie is generally a rock-ledge or jutting crag upon which a platform of sticks, grass, moss, etc., is formed and added to year by year. Eggs: 2 or 3; laid in April; greyish-white, variably mottled, streaked or blotched with rust-colour and with underlying grey markings; size 2·90 by 2·30. It is addicted to lamb-stealing, but chiefly preys on mountain-hares, ptarmigan, grouse, etc.

GENUS XC. HALIAËTUS, *Savigny (1810).*

Bill much as in *Aquila*, but cutting edges of upper mandible exhibiting an indistinct tooth. Tarsus feathered on upper half only, anterior portion of lower tarsus and all upper surface of toes covered with broad scales.

168. Haliaetus albicilla (Linn.). WHITE-TAILED EAGLE.

Hab. Palæarctic region; also south of Greenland.

Male: head and neck greyish-white; tail pure white, slightly graduated; whole remainder of plumage dark brown, becoming slate-black on wing-quills; bill yellowish with a darker tip; cere, irides and feet yellow; claws blackish. Length 28·00 to 29·00. Female: length about 33·00, otherwise similar. Young: head and tail dark brown like rest of plumage; bill dusky; iris hazel. Even after white tail is acquired head remains light greyish-brown, the greyish-white head probably only being obtained by very old birds.

The present breeding range of this eagle is practically the same as that of last species. It is, however, the more common of the two as a visitor to England, immature examples occurring with some regularity on the east side and being (from their dark colour) recorded with equal regularity in the newspapers as "Golden Eagles." Nesting habits are much like those of *A. chrysaetus;* eggs are dull white, usually unmarked; size 2·85 by 2·25.

GENUS XCI. ASTUR, *Lacépède (1801).*

Bill rather short, upper mandible decurved from base, its cutting edges showing a semi-obsolete tooth. Wings somewhat short. Tarsus moderate, rather stout, broadly scaled; claws long, much curved.

169. Astur palumbarius (Linn.). GOS-HAWK.

Hab. Palæarctic region. In winter partially migratory.

Until within a hundred years ago this fine hawk bred in Scotland but it is now a rare straggler to any part of the British Isles.

170. **Astur atricapillus (Wilson).** AMERICAN GOS-HAWK.

Hab. Northern Nearctic region. Partially migratory in winter.

One example has been killed in Perthshire (1869), a second in Tipperary (1870), and a third in King's Co. (1870).

GENUS XCII. ACCIPITER, *Brisson (1760).*

Bill with cutting edges of upper mandible festooned, forming a rounded and not very prominent tooth. Tarsus rather long, slender, the lower portion covered with short scales, the rest smooth. Wings short.

171. **Accipiter nisus (Linn.).** SPARROW-HAWK.

Hab. Whole Palæarctic region.

Male: above dark bluish-slate; wing-quills being nearly black, and tail crossed by several broad bands of black and narrowly tipped with greyish-white; ear-coverts reddish; nape mottled with white: below white, barred transversely with pale reddish-brown; bill bluish-black; cere, irides and tarsi yellow; claws black. Length 12·00. Female: rather browner above; under parts barred with dark brown; old birds often resemble male except in being considerably larger; length 14·50.

Fairly common everywhere in woodlands, but rarer in more open country. Nest: always situated in a tree, usually in forks of the main limbs from 20 to 30 feet up; undoubtedly that of a Crow, Magpie or Ring-dove is often utilized and added to, yet in spite of recent assertions to the contrary ("Ornithologist," May, 1896), I believe that this species more often than not builds its own nest. Eggs: 4 to 6; bluish-white, handsomely blotched with two shades of reddish-brown, often in a zone around larger end; size 1·60 by 1·28. A bold and spirited hunter, often making havoc among young game, chickens and ducklings, but usually preying upon small birds; it devours its food on the ground.

GENUS XCIII. MILVUS, *Cuvier (1800).*

Bill with upper mandible decurved from middle only, its cutting edges nearly straight. Wings and tail long, the latter considerably forked. Feet stout; tarsus short.

172. Milvus milvus (Linn.). KITE.

Hab. Europe, except extreme north and east.

Male: head and nape greyish-white with dusky streaks; remaining upper parts and tail reddish-brown with blackish centres to feathers of mantle; wing-quills blackish; below rust-colour with blackish striations; iris sulphur-yellow; cere and tarsi yellow; claws black. Length 24·50. Female: plumage similar; length 26·00.

As a breeding species it has now probably ceased to exist in England; although in several localities in Wales (and possibly also in Scotland) a few pairs still breed (1896), and although the nest is taken more often than not one or two broods seem to be reared safely. In Ireland five or six birds have been observed, but only one taken.

173. Milvus migrans (Bodd.). BLACK KITE.

Hab. Europe, south of the Baltic. In winter migrating to Africa.

An example in the Newcastle Museum was taken at Alnwick, Northumberland, 11th May, 1866, and recorded in the Ibis (1867, p.253) by the late John Hancock.

GENUS XCIV. ELANOIDES, *Vieillot (1823).*

174. E. furcatus (Linn.). SWALLOW-TAILED KITE.

Hab. Tropical and temperate America, northward (locally) to northern United States. In winter migratory.

Mr. Harting (Handbook, p. 88) gives the records of this species as: one in Argyleshire (1772); one in Yorkshire (1805); one on the Mersey (1843); one in Cumberland (1853); one in Surrey (1856).

GENUS XCV. PERNIS, *Cuvier (1817)*.

Bill moderate, somewhat slender, upper mandible decurved nearly from the base and with its cutting edges almost straight. Tail rather long, slightly rounded. Feet tolerably stout; tarsus feathered on upper part; rest of foot reticulated; claws very moderately curved.

175. Pernis apivorus (Linn.). HONEY-BUZZARD.

Hab. Palæarctic region, in Europe north nearly to Arctic Circle. In winter migrating to Africa.

Male: head and nape grey; remaining upper plumage dark brown, the tail having several dusky bars; below whitish, spotted on throat and barred on breast and sides with brown; iris, cere and feet yellow. Length 22·00. Female: head and nape brown. Length 24·00.

A summer visitor. It has bred during the last century in nearly every part of England and even in Scotland, the last known instances having been in Hampshire. To Ireland it is a rare straggler, the only recent occurrences recorded being in 1890 and 1892 (Co. Wexford).

GENUS XCVI. HIEROFALCO, *Cuvier (1817)*.

A sub-genus of *Falco* containing the group of Gyr-Falcons which inhabit the circumpolar regions.

176. Hierofalco gyrfalco (Linn.). GYR FALCON.

Hab. Arctic Scandinavia.

One example, shot in Sussex, January, 1845, has been identified by Mr. Gurney, and a second, shot in Suffolk,

October, 1867, is figured in Babington's "Birds of Suffolk," and was also examined by Seebohm. This is the true *F. gyrfalco* of Linnæus, closely allied to *H. islandus*, but smaller and darker.

176a. Hierofalco gyrfalco islandus (Gmel). ICELAND FALCON.

Hab. Iceland and South Greenland. Wandering southward in winter.

Although this form is included in the B. O. U. List as an apparent *species*, all the *Grey* Falcons are too closely related to be specifically separated, and the Iceland Falcon must therefore be classed as a sub-species of *H. gyrfalco*, since the latter was the earliest-described form. The present race is distinguished from *H. candicans* by having general tint of upper parts deep grey, with short blackish bars or spots on all the feathers; tail with 8 or 10 transverse bars; under parts greyish-white, with longitudinal dark spots and short bars on the flanks; bill lead colour; iris deep brown; cere and feet bluish-grey. Length 21·50. Female: similar in plumage. Length 24·00.

A rare winter visitor, but it has occurred at long intervals in most parts of the British Isles.

177. Hierofalco candicans (Gmel). GREENLAND FALCON.

Hab. Greenland and Arctic America. Wandering southward in winter.

Readily distinguishable from the Grey Falcons owing to the fact that ground tint of upper as well as lower parts is *white* at all ages; the upper parts are marked much as in the Iceland Falcon, but the lower parts are much more sparingly spotted; cere and tarsi appear to be yellowish in old birds, but in younger examples these and the irides are as described for Iceland Falcon, the length of adults also being usually nearly the same as there given. Like the last a rare winter visitor.

Genus XCVII. FALCO, *Linnæus (1766).*

Bill rather short, strong, upper mandible decurved from base, its cutting edges each side with a sharp and strong tooth. Wings long, pointed. Tarsus short, strong; toes long; claws much curved, sharp.

178. Falco peregrinus, Tunstall. Peregrine Falcon.

Hab. Northern Palæarctic region; also Greenland.

Male: cap and sides of head dull black, as is also a moustache-like stripe running backward from the gape; remaining upper parts bluish-slate, with dusky bars on the mantle; tail with several bands of black; throat white, tinged with buff and sparingly striped with black; remaining under parts whitish, with dusky bars; bill bluish-black; iris deep brown; cere and feet yellow; claws black. Length 15·50. Female: similar in plumage. Length 18·00. Young: feathers of upper parts brown, with pale margins; under parts with a yellowish tint and dusky streaks; cere and tarsi brownish.

Still breeds sparingly along west side of Great Britain, especially in Wales and Scotland; also in many parts of Ireland. Eggs: 2 to 4; usually laid on a ledge or in a fissure in high cliffs; if a nest of another bird is not appropriated they are deposited in a hollow in the scanty soil; ground tint varies from yellowish to pale rufous, closely spotted with reddish-brown of two shades; size 2·00 by 1·60. The same nesting site is resorted to for many successive years. It is the falcon of mediæval falconry, and is a magnificent hunter.

179. Falco subbuteo, Linn. Hobby.

Hab. Palæarctic region. In winter moving southward.

Male; above dark slate; throat and sides of neck white, with a black stripe on each side running from the gape; below yellowish-white, boldly streaked with black; thighs and under tail-coverts chestnut; bill lead-colour; iris brown; cere and tarsi yellow; claws black. Length 12·00. Female: plumage similar. Length 13·50. Young: feathers above brown, with buff margins; thighs and lower tail-coverts merely tinged with rust-colour.

A regular but scarce summer visitor, arrives in May leaving by October; has bred at intervals in all our southern and eastern counties, but rarely straggles to Scotland; has occurred eight times in Ireland. Eggs seem to be almost invariably laid in the deserted nests of Crows, etc., built in trees in wooded districts; they are similar to but often yellower in ground tint than those of Kestrel; size 1·58 by 1·25. Nestlings are clothed with whitish down, the back, however, being dark grey, while nestling Kestrels are covered with purely white down.

180. Falco æsalon, Tunstall. Merlin.

Hab. Northern Palæarctic region. In winter partially migratory.

Male: feathers above bluish-slate with narrow black shaft-streaks, excepting the neck which is reddish-brown; tail tipped with white and with a broad sub-terminal band of black; upper throat whitish; rest of under parts pale reddish-brown with dusky-streaks; iris brown; cere and tarsi yellow; claws black. Length 10·25. Female: above deep brown, the tail with several buffish bars; under parts whitish with brown streaks. Length 12·00.

Breeds on moorlands of Scotland and northern half of England; also in Wales and mountain districts of Ireland. Eggs (4 or 5) are laid on the ground; smaller than those of Kestrel or Hobby; size 1·50 by 1·18. Old nests in trees are occasionally made use of, but instances of this proceeding are rare. Preys on birds up to size of a thrush.

GENUS XCVIII. TINNUNCULUS, *Vieillot (1807)*.

A sub-genus of *Falco*, containing in all a dozen species.

181. Tinnunculus vespertinus (Linn.). RED-FOOTED FALCON.

Hab. Central and Eastern Europe; also Western Asia. In winter migrating to Africa.

Male: whole plumage of a deep sooty slate-colour, excepting the thighs and under tail-coverts which are chestnut; iris dark red; cere and tarsi orange-red; claws horn-white. Length 11·00. Female: mantle, wing-coverts and tail slate-grey with dusky bars; head, nape, and under parts rufous-brown. Length 12·00.

A not uncommon visitor in spring or autumn; examples been taken in almost every county in the southern half of England, but more rarely in the north. It has occurred but three times in Scotland and once in Ireland (1832).

182. Tinnunculus tinnunculus (Linn.). KESTREL.

Hab. Palæarctic region; also Africa and British India. Partially migratory in winter.

Male: mantle bright rufous spotted with black; rest of upper parts bluish-grey; the tail paler, broadly tipped with white and with a wide black sub-terminal band; under parts rusty-buff with blackish spots and streaks; circumocular region black; bill lead-coloured; iris deep brown; cere and tarsi yellow; claws black. Length

12·75. Female: feathers above dull reddish-brown barred with black; below similar to male; tail rufous, barred with black and with a white tip and black sub-terminal band. Length 14·25. Young: like female.

Quite the commonest of the *Accipitres* with us. In wooded districts the 5 or 6 eggs are laid in old nests of the Crow, but in mountainous parts they are deposited on the bare soil on rock-ledges or in crevices; they are yellowish-white, clouded and closely spotted all over with rusty-red; size 1·60 by 1·30. Food consists largely of mice with some insects, but I have known it to take small birds in winter.

183. Tinnunculus cenchris (Naum.). LESSER KESTREL.

Hab. Southern Europe. In winter migrating as far as South Africa. A straggler to North Europe.

Three occurrences are known, *i.e.*, one near York (1867), one near Dover (1877), and one in Co. Dublin (1891).

GENUS XCIX. PANDION, *Savigny (1810).*

Bill stout and strong; tooth nearly obsolete. Feathers of crown forming a short crest. Wings very long and narrow. Feet stout and strong; tarsi short, reticulated; outer toe reversible; claws large, sharp, much curved.

184. Pandion haliaëtus (Linn.). OSPREY.

Hab. Palæarctic region.

Male: crest whitish streaked with dark brown; rest of upper parts deep brown; below whitish with an irregular brown breast band; bill blackish; iris yellow; cere and feet bluish; claws black. Length 21·00. Female: brown gorget more pronounced and crest more streaked; also larger. Length 23·00. Young: feathers of upper parts with pale margins and crest showing very little white.

A few birds still breed on the lochs in certain parts of the Highlands, but to rest of British Isles it is merely a visitor (but by no means a rare one) in the autumn and winter months. It preys entirely upon fish.

ORDER STEGANOPODES.

Family Pelecanidæ.

GENUS C. PHALACROCORAX, *Brisson (1760).*

Bill moderately long, straight, compressed, upper mandible much hooked at tip. Nostrils basal, concealed. Face and chin naked. Wings and tail moderate. Feet placed far back; tarsus short; three toes in front, one on inner side of tarsus, all united for whole length by a web.

185. Phalacrocorax carbo (Linn.). CORMORANT.

Hab. Palæarctic region and Eastern Nearctic region.

Adult: crown and nuchal crest black, with a number of narrow white feathers interspersed; feathers of back and wing-coverts deep brown, edged with black, and with glossy reflections; wing-quills and tail black; upper throat and region behind eye white; entire under-side bluish-black, excepting a patch of white on each flank; bill pale brown; bare skin of chin which forms a slight pouch, yellow; iris green; feet and webs blackish. Length 34·00. In autumn slightly lengthened crest-feathers are lost, as are also white patches on flanks. Young: above brown; below mottled with white and pale brown; iris brown.

Found breeding (usually in colonies) on all our rocky coasts. Nest: usually on rock-ledges, and composed of a pile of sea-weed, sticks and grass. Eggs: often 3; shell pale blue and fairly smooth, but concealed by a chalky incrustation which can be scraped off with a knife; size 2·70 by 1·60. Feeds entirely on fish.

186. Phalacrocorax graculus (Linn.). SHAG.

Hab. Western Europe.

Adult: plumage glossy blackish-green; tail of 12 feathers instead of 14 as in *P. carbo*; spring crest situated on fore-part of head and recurved forward; bill blackish, under mandible yellow at base; bare skin of gape black, speckled with yellow; iris green; feet and webs black. Length 26·00. Young: above greenish-brown; below greyish-brown, mottled with darker brown.

Found on all the most rocky coasts of the British Isles; it is much more common than the preceding species in the Shetland, Orkneys, and Hebrides. Nesting habits and eggs similar, but latter are smaller; size 2·45 by 1·45.

GENUS CI. SULA, *Brisson (1760).*

Bill long, straight, of a lengthened and slightly compressed conical shape, upper mandible very slightly decurved towards the point. Face and throat naked. Wings long; tail moderate, pointed.

187. Sula bassana (Linn.). GANNET.

Hab. Coasts of the North Atlantic.

Adult: whole plumage white, excepting the primaries which are black; head and neck tinged with warm buff. Length about 32·00. Young: above deep brown, mottled with white; below mottled with dark grey and pale brown; full adult dress is not assumed until the bird is four or five years old.

Breeds on the Welsh Coast and also on Lundy Island, but nowhere in the east of England; there are many large colonies around the Scotch coasts, and two off the south-west coast of Ireland. Nest: a pile of seaweed, grass, etc. The single egg is bluish-white but covered with a white incrustation which speedily soils; size 3·10 by 1·95. Food consists of such fish as may be visible near the surface of the water, the bird descending upon them with great rapidity from some height.

ORDER HERODIONES.

Family, Ardeidæ.

GENUS CII. ARDEA, *Brisson* (1760).

Bill long, straight, compressed, stout at base, and diminishing to an acute point; head crested. Wings ample, of moderate length. Tail of 12 feathers, short; lower part of tibia bare; tarsus long, slender, scaled in front; toes three before, one behind.

188. Ardea cinerea, Linn. COMMON HERON.

Hab. Palæarctic and Oriental regions.

Male: forehead and cheeks white; long filamentous, non-erectile crest blackish-blue; upper plumage bluish-slate; primaries black; front of neck white, marked down centre with longitudinal rows of dark slate-blue spots, the feathers terminating on breast in long plumes; below greyish-white, with some dusky streaks; bill yellowish; iris yellow; tarsi and toes yellowish-green; claws black. Length nearly 36·00. Female: similar, but crest and breast plumes are shorter.

Common throughout British Isles, usually nesting in the tree tops in tolerably large protected heronries, but in wilder parts nests may be found on cliffs and rocks as commonly as on trees. Nest: large and flat, of sticks, lined with grass, moss, and roots; often re-used and added to yearly. Eggs: 3 to 5; pale bluish-green; size 2·50 by 1·70. Food: frogs, small reptiles, mollusca, eels, and other fish, also water rats and even young water-fowl. The slow steady movement of its hollowed wings as it passes over at some height readily identifies it; at intervals it utters a deep, hoarse, barking note.

189. Ardea purpurea Linn. PURPLE HERON.

Hab. Temperate and tropical regions of eastern hemisphere; in Europe north in summer to Holland.

Adult is, to some extent, similar to *A. cinerea*, but forehead and centre of crown are purplish-black, like crest; long plumes, which overlap wings when folded, reddish-brown instead of grey like rest of upper parts; cheeks and neck reddish-buff with a stripe of spots on each side; lower breast and under wing-coverts rich chestnut-red. Length 32·00.

Has occurred more than forty times in England, three times in Scotland, and once in Ireland.

190. Ardea alba, Linn. GREAT WHITE HERON.

Hab. Southern Palæarctic region. Migrates southward in winter.

Not more than ten examples have been taken in Great Britain, most of them many years ago, while in Ireland it has never been obtained. Several more are said to have been observed at long intervals.

191. Ardea garzetta, Linn. LITTLE EGRET.

Hab. Southern Palæarctic and Oriental Regions; also Africa. In winter migrates from more northern districts.

An even rarer visitor than the last. Nine examples are said to have been taken in England, but some of these occurrences are doubtful. One is said to have been killed in Scotland in 1844, while Thompson records it as obtained three times in Ireland.

192. Ardea bubulcus, Audouin. BUFF-BACKED HERON.

Hab. South-Western Europe and whole of Africa.

An example in the Natural History Museum was killed in Devon in 1805. A second is said to have been taken in Devon in 1851 (Zool. 1851, p. 3116), while Stevenson records one taken near Yarmouth in 1827, but neither of these exist.

193. Ardea ralloides, Scop. SQUACCO HERON.

Hab. Southern Europe and Africa. Migrating from Europe in winter. A straggler to North Europe.

Adult: plumage above and on front of neck buff, head and back of neck being marked with dusky striations; crest-feathers white in centre with blackish margins; under parts white; bill bluish at base, black at tip; iris yellow; feet orange. Length about 18·00. Immature birds have colours less pure and upper parts principally brown.

Has occurred nearly forty times in England, twice in Scotland, and about seven times in Ireland.

GENUS CIII. NYCTICORAX, *Stephens (1819)*.

Differs to no very great extent from *Ardea*.

194. Nycticorax griseus (Linn.). NIGHT-HERON.

Hab. Southern Palæarctic region; also Africa.

Male: above glossy greenish-black, excepting back of neck and wings which are slate-grey; filamentous crest-feathers white; below greyish-white; bill blackish above, blue-grey below; iris red; tarsi yellow. Length about 21·00. Female: similar, but has less crest. Young: chiefly dark brown with whitish spots and pale buff streaks.

A tolerably regular visitor to southern coasts of England where it occurs almost annually in spring and occasionally in autumn. Northward it is rare and has only occurred five or six times in Scotland. In Ireland about a dozen have been taken, the last in October, 1893.

GENUS CIV. ARDETTA, *Gray (1842)*.

Differs materially from last two genera in having tail of *ten* short soft feathers. Tarsus of moderate length only; tibia feathered to tarsal joint.

195. Ardetta minuta (Linn.). LITTLE BITTERN.

Hab. Europe, south of about lat. 55°; also Northern Africa and South-West Asia. In winter migrating from Europe.

Male: crown, nape and mantle glossy greenish-black; wing-coverts buffish-white; primaries and tail blackish-brown; sides of head and front of neck rufous-buff; under parts paler buff, marked with dusky streaks on breast and flanks; bill and feet greenish-yellow. Length 12·50. Female: mantle dark brown; wing-coverts pale brown; sides of head and back of neck reddish-brown; under parts much more streaked. Length 12·00.

An irregular summer visitor to east and south of England; it is generally supposed that it may have bred in Norfolk and elsewhere, although the nest has never been found. To Scotland it is a rare straggler. In Ireland about twelve occurrences have been recorded.

GENUS CV. BOTAURUS, *Stephens (1819)*.

Bill and feet much as in *Ardea*; tibia bare on lower portion; tail of *ten* short soft feathers.

196. Botaurus stellaris (Linn.). BITTERN.

Hab. Palæarctic region; also whole of Africa. In winter migrating from northern regions.

Male: plumage buff, barred and pencilled above and streaked below with brownish black; primaries barred with rufous buff and black; crown and nape black; feathers of neck and upper breast long and distended forming a ruff; bill greenish-yellow; tarsi and toes dark green. Length 27·00. Female identical.

Formerly a common breeding species in many parts of British Isles, but now chiefly a casual spring visitor. It is tolerably certain that it continued to breed in Norfolk until within ten years ago, while in the "Midland Naturalist" for April, 1885, appear the particulars of a nest found on a large pool at Sutton Park, Warwickshire, in 1884.

197. Botaurus lentiginosus (Montagu). AMERICAN BITTERN.

Hab. North America. In winter migrating southward.

A not uncommon straggler to the British Isles, principally during the autumn and winter. In England and Wales it has occurred about nine times; in Scotland four times; and in Ireland nine times. It is slightly smaller than *B. stellaris* and darker in colour, upper parts being much more closely and finely marked; primaries also are *uniform leaden-brown* instead of being barred as in *B. stellaris*.

Family Ciconiidæ.

GENUS CVI. CICONIA, *Brisson (1760).*

198. Ciconia alba, *Bechstein.* WHITE STORK.

Hab. Southern Palæarctic region; in Europe north to Southern Scandinavia. In winter migrating southward.

A scarce and irregular spring visitor to East Anglia, although never known to breed; it has also been recorded from several southern counties, but very few have reached Scotland; in Ireland an example was taken in Co. Cork in 1846, and another in 1866.

199. Ciconia nigra (Linn.). Black Stork.

Hab. Southern Palæarctic region; also Africa. In winter migrating from Europe.

A very rare straggler in summer to south and east of England, where thirteen examples in all have been taken.

Family Ibididæ.

Genus CVII. PLEGADIS, *Kaup (1829).*

Bill very long, slender, decurved, stout at base, tip abrupt and rounded, upper mandible grooved; nostrils in upper part of bill, near base, slit in a membrane. Feet much as in *Platalea.*

200. Plegadis falcinellus (Linn.). Glossy Ibis.

Hab. Southern Palæarctic and Oriental regions; also Eastern United States. Migrating southward in winter.

Apparently a not uncommon visitor a century ago, but is now of decidedly rare occurrence; has, however, occurred in autumn at long intervals in all parts of the British Isles, although most frequently in east and south of England.

Family Plataleidæ.

Genus CVIII. PLATALEA, *Linnæus (1766).*

201. Platalea leucorodia, Linn. Spoonbill.

Hab. Palæarctic and Western Oriental regions. In winter migrating southward.

Two hundred years ago this species bred commonly in England, but it is now merely a scarce and irregular visitor to the eastern and southern counties; rarer in the west and north, while very few stragglers have been taken in Scotland or Ireland.

ORDER ODONTOGLOSSÆ.

Family Phœnicopteridæ.

GENUS CIX. PHŒNICOPTERUS, *Brisson (1760).*

202. Phœnicopterus roseus, Pallas. FLAMINGO.

Hab. Southern Europe, whole of Africa, and West Asia.

One was shot in Staffordshire in September, 1881, and a second in Hampshire, November 26th, 1883; another was observed by Captain Shelley at New Romney, August 12th, 1884 (Saunders, Manual, p. 383).

ORDER ANSERES.

Family Anatidæ.

GENUS CX. ANSER, *Brisson (1760).*

Bill moderate, stout and high at base, which is furnished with a cere, and terminating in a hard "nail"; edges of mandibles furnished with transverse plates or laminæ. Nostrils at about middle of upper mandible. Wings ample, moderately long. Tail short, of 16 feathers. Tarsus moderate, stout; three front toes united for whole length by a web; hind toe small, elevated.

203. Anser cinereus, Meyer. GREY LAG GOOSE.

Hab. Palæarctic region. In winter migratory.

Adult: head, nape and mantle greyish-brown; rump and outer portion of all the wing-coverts light bluish-grey; upper tail-coverts white; throat and breast grey; under parts white, with dusky spots on the sides; bill flesh-coloured, with a white nail; iris brown; tarsi, toes and webs flesh-colour; claws blackish. Length 33·00. Female: plumage similar; length about 30·00.

Much less common than formerly, and now only breeding in the extreme north of Scotland and in the Hebrides; to the rest of Great Britain and to Ireland it is a winter visitor only, although in the latter, a colony of semi-domesticated birds has existed for many years on a lake at Castlecoole, Co. Fermanagh. In Scotland nest is placed on ground among heather, rushes, or rank grass, and is lined with down from the bird's breast. Eggs: 5 to 8; creamy-white; size 3·45 by 2·35.

204. Anser segetum (Gmel.). BEAN-GOOSE.

Hab. Northern Palæarctic region. In winter migratory.

Slightly smaller than the last, but more slender and with a longer wing; its appearance is considerably darker, and the blue-grey of shoulders is absent; bill is orange with the base black and a *black nail*; tarsi, toes and webs are orange. A winter visitor to most parts of British Isles.

205. Anser brachyrhynchus, Baillon. PINK-FOOTED GOOSE.

Hab. Arctic regions, breeding in Iceland, Spitsbergen, and probably other areas. In winter migrating to temperate Europe and Asia.

Closely allied to the last, and like it the base, edges, and nail of bill are black, but centre part is pink instead of orange; tarsi, toes, and webs also pink; tail feathers more broadly edged with white, and shoulder showing some bluish-grey, of a darker tint, however, than in *A. cinereus.* It is still smaller than *A. segetum*, length of male being 28·00 to 29·00; female slightly smaller.

A regular winter visitor to the east coast, but rarer in the west or south. In Ireland, where the last species is of regular occurrence, the present form has only once been obtained, *i.e.*, an example shot on Lough Swilly in October, 1891 (Zool., 1892, p. 33).

206. **Anser albifrons** (Scop.). White-fronted Goose.

Hab. Northern Palæarctic region (except Scandinavia), breeding chiefly within the Arctic Circle. In winter migrating southward.

Closely related to the three preceding species, but adult is at once distinguished by its having the feathers on forehead and around base of upper mandible white; breast and sides are broadly banded with black and bill is orange, but with a *white* nail; feet are also orange. Length from 26·00 to 28·00, female being slightly smaller than male. A regular winter visitor, but apparently most common on the west side. It is the commonest of the "Grey Geese" which visit Ireland.

206a. **A. albifrons erythropus** (Linn.). Lesser White-fronted Goose.

Hab. Arctic Norway. In winter migrating southward.

A smaller race than the last; distinguished chiefly by having the white on forehead extending fully as far behind as the level of the eye; bill also is shorter, and ridge above is straight, forming a line with the forehead. A genuine example was shot by Mr. A. C. Chapman at Holy Island, Northumberland, September 16th, 1886.

Another, shot in Somerset in January, 1888, was identified by Mr. Cecil Smith, but he appears to have been doubtful whether it was an escaped bird or not.

GENUS CXI. CHEN, *Boie (1822).*

207. Chen hyperboreus (Pallas). LESSER SNOW GOOSE.

Hab. Western Arctic America and N.E. Asia. In winter migrating southward; straggling to British Isles.

Three examples were shot in Wexford in 1871; two more taken alive from a flock of seven in Co. Mayo in 1877; one observed by the Rev. H. A. Macpherson in Cumberland in 1884; and one shot in Co. Mayo in 1886; finally the species occurred in small flocks in Northumberland, Yorks and Cumberland in winter of 1890—1.

In B. O. U. List it bears the name of *C. albatus* (Cassin), but it has since been shown that this must give place to Pallas's earlier name (*hyperboreus*) conferred on specimens obtained in Asia. A larger sub-species inhabits *Eastern* Arctic America, but none of British specimens have been referred to this race.

GENUS CXII. BRANTA, *Scopoli (1769).*

Bill considerably shorter and narrower than in *Anser;* not nearly so long as head, and higher than wide at base.

208. Branta bernicla (Pallas). BRENT GOOSE.

Hab. Northern Palæarctic and North-eastern Nearctic regions, breeding within Arctic Circle. In winter migratory.

Male: head, neck and upper breast black, except a small white semi-collar on each side of neck; mantle feathers blackish-brown with paler margins; rump black;

upper and under tail-coverts white; tail and wing-quills black; belly and sides dark slate; bill black; feet brownish-black. Length 23·00. Female: similar but slightly smaller.

A common winter visitor to all parts of the British coasts, but particularly the eastern side; also frequently seen inland; it is quite the commonest of our wild geese.

208a. Branta bernicla glaucogaster (Brehm).
WHITE-BELLIED GOOSE.

Hab. Arctic regions. In winter migrating southward.

Distinguished from the typical bird by having the belly whitish. It occurs in our islands as a winter visitor.

209. Branta leucopsis (Bech.). BERNICLE GOOSE.

Hab. Arctic regions of Old World and Greenland; probably breeding very far north. In winter southward.

Readily distinguished from *B. brenta* by its having forehead, sides of head and chin white, excepting a black patch on the lores; mantle pale grey barred with black, and abdomen paler; otherwise much the same but larger.

A common winter visitor, chiefly to the west side and to Ireland.

210. Branta ruficollis (Pallas). RED-BREASTED GOOSE.

Hab. Eastern Siberia.

Has occurred five times on the east coast of England, also twice in Devon and once in Scotland.

GENUS CXIII. CYGNUS, *Bechstein (1809).*

Bill about as long as head, higher than broad at base, depressed towards tip. Neck very long, slender.

211. Cygnus olor (Gmel.). MUTE SWAN.

Hab. Palæarctic region.

Common everywhere in a semi-domesticated state while in many localities numbers of birds which have reverted to a wild state are found. It is distinguished from other species by the protuberance at base of upper mandible.

211a. C. olor immutabilis (Yarrell). POLISH SWAN.

As this form holds specific rank in the B. O. U. list, I have allowed it to stand here, tentatively, as a sub-species of the last. Many authorities consider it to be merely an abnormal variety of *C. olor*, in which the cygnets are usually dark grey, while in the Polish Swan they are white. Birds referred to this variety are occasionally shot on our east coast in winter.

212. Cygnus musicus, Bech. WHOOPER SWAN.

Hab. Northern Palæarctic region.

Male: white; basal three-fifths of bill pale yellow, this colour extending below and *beyond the nostrils*; rest of bill and the feet black. Length 58·00. Female: slightly smaller. Cygnet: greyish-brown; bill flesh-coloured with the edges and tip black; feet flesh-coloured.

A fairly common visitor during severe weather.

213. Cygnus bewicki, Yarrell. BEWICK'S SWAN.

Hab. Eastern Arctic Europe and Arctic Asia.

Male: white; bill black, excepting an oval patch of orange-yellow covering one-third of basal part of upper mandible each side, but falling much short of the nostrils; feet black. Length 48·00. Female: slightly smaller.

A common visitor to our coasts during severe winters.

GENUS CXIV. TADORNA, *Flemyng (1822)*

Bill as long as head, elevated at base and furnished with a protuberance or knob, much depressed in middle; nail decurved.

214. Tadorna tadorna (Linn.). COMMON SHELD-DUCK.

Hab. Palæarctic regions.

Male: head and upper neck deep glossy green; lower neck white; upper mantle, shoulders and breast chestnut: wing-coverts white, "speculum" metallic green, rest of wing blackish-brown; remainder of upper parts white with a black tip to tail; middle of belly deep brown, sides white; bill and protuberance bright red; iris reddish-brown; feet flesh colour. Length 24·00. Female: lacks protuberance on bill. Young: neck and head brown, with a good deal of white on latter; whole under parts white; bill flesh-colour; feet purplish-grey.

Breeds sparingly on sandy shores on all parts of the British coasts. Nest: commonly placed in rabbit burrows, occasionally under rocks or in a hole excavated by the bird; grass and bents, lined with down from the bird's breast. Eggs: 8 to 12; creamy-white; 2·70 by 1·90.

GENUS CXV. CASARCA, *Bonaparte (1838)*.

215. Casarca casarca (Linn.). RUDDY SHELD-DUCK.

Hab. Southern Palæarctic region.

Prior to 1892 a very rare or even doubtful visitor, only 9 or 10 occurrences being on record, but in the year mentioned a considerable immigration took place, nearly 20 examples being shot, chiefly from small flocks, on various parts of our coasts.

Genus CXVI. MARECA, *Stephens (1824)*.

Bill shorter than head, higher than wide at base, depressed and narrowing towards tip; laminæ scarcely discernible. Tarsus short; hind toe possessing a narrow lobe.

216. Mareca penelope (Linn.). Wigeon.

Hab. Northern Palæarctic region; also Alaska and Greenland. In winter migratory.

Male: crown and forehead cream-colour; rest of head and neck chestnut, speckled on cheeks and nape with dark green, and with a stripe of same through eye; mantle and scapulars greyish-white, with fine dusky vermiculations; wing-coverts white, tipped with black; speculum metallic green followed by a black band; primaries and tail blackish; breast and belly white; under tail-coverts black; sides greyish with fine dusky vermiculations; bill bluish-slate; iris dark brown; feet dark brown. Length 17·50. Female: slightly smaller; head and neck reddish-brown, spotted with darker brown; feathers of upper parts dusky in centre with pale margins; wing-coverts whitish; speculum greyish-green; breast pale brown; under white mottled with buff. About August male assumes for a time a similar plumage to female's, this being also the case in most succeeding genera.

A common winter visitor; in Scotland also breeding throughout the Highlands and more sparingly in the Orkneys and Shetlands.

The down-lined nest is concealed in heather, rushes or rank grass. Eggs: 7 or 8; cream-colour; 2·20 by 1.50.

217. Mareca americana (Gm.). American Wigeon.

Hab. Western Arctic America. In winter southward to Central America and West Indies.

An example was obtained by Blyth in the winter of 1837—38 in Leadenhall Market. This specimen is still in existence, but not so some others said to have been obtained since in various localities.

GENUS CXVII. DAFILA, *Stephens (1824).*

Tail pointed, middle feathers much lengthened in the male. Neck rather long, slender. Bill much as in *Mareca*.

218. Dafila acuta (Linn.). PINTAIL.

Hab. Northern Palæarctic and Nearctic regions. In winter southward nearly to Equator.

Male: head and throat glossy dark reddish-brown; darker on back of neck, which has a white stripe down each side; wing-coverts dull buff; speculum green, followed by a band of black and another of white; other plumage much as in *Mareca penelope*; bill dusky, edged with bluish-grey; iris dark brown; feet slate-brown. Length 26·00 while long tail-feathers appear. Female: mottled with dark and pale brown above; speculum green; below nearly white, slightly speckled on neck and breast; tail short, dark brown with whitish bars.

A fairly common winter visitor; it breeds in the Hebrides and probably also in the Orkneys.

GENUS CXVIII. ANAS, *Brisson (1760).*

Bill as long as head, depressed and somewhat broad; laminæ apparent on both mandibles. Hind toe small, elevated, unlobed.

219. Anas boscas, Linn. WILD DUCK.

Hab. Palæarctic and Western Nearctic regions.

Male: Length 22·00. Female: feathers of whole plumage chiefly having dusky centres and pale brown margins, producing a mottled appearance; central tail-feathers not curled up as in male; speculum of wing dark metallic green. Length 20·50.

Breeds more or less commonly throughout the British Isles; very abundant during winter. Nest: on ground near water; in hedge-bottoms or concealed in rank grass, rushes, brambles, or growing corn. Eggs: 8 to 10; buffish white, tinged with green; 2·25 by 1·65. Feeds at night; food comprising mollusca, aquatic insects, etc., as well as vegetable matter.

Genus CXIX. CHAULELASMUS, *Gray (1845).*

Chiefly characterized by the laterally projecting laminæ of the upper mandible of bill.

220. Chaulelasmus streperus (Linn.). Gadwall.

Hab. Palæarctic and Nearctic regions.

Male: head and neck pale brown with darker spots; feathers of back and lesser wing-coverts dark in centre with pale grey margins; median coverts rufous, greater blackish; *speculum white;* primaries and tail dark brown; rump and both upper and lower tail-coverts bluish-black; feathers of upper breast and flanks dusky grey with paler margins; lower breast and belly white; bill bluish-slate; iris hazel; feet orange. Length about 20·00. Female: feathers of upper parts and breast margined with pale brown instead of grey; *speculum white.*

Breeds rather commonly in Norfolk, but in the rest of British Isles is only known as a scarce winter visitor.

Genus CXX. QUERQUEDULA, *Stephens (1824).*

Bill nearly as long as head, depressed and not broad; the laminæ visible but not projecting.

221. Querquedula querquedula (L.). GARGANEY.

Hab. Palæarctic region, except extreme north. In winter migrating to North Africa and India.

Male: forehead, crown and back of neck dark brown, bordered each side by a white stripe which commences above eye and fades half-way down neck; cheeks and throat rufous-brown, minutely streaked with white; chin blackish; breast-feathers pale brown with dusky margins; back dark brown; wing-coverts light bluish-grey; speculum metallic-green, bordered above and below with white; abdomen white; sides with fine dusky vermiculations, and with two noticeable black crescents on each lower flank; bill blackish; iris hazel; feet slate-brown. Length 15·50. Female: white stripe on side of head less distinct and tinged with buff; wing-coverts brownish-grey; speculum duller; chin and under parts greyish-white, mottled with two shades of brown on breast and sides.

A scarce visitor in spring; seldom found breeding out of East Anglia, where it nests regularly in small numbers. Of uncommon occurrence in Scotland and a very rare straggler to Ireland.

222. Querquedula discors (L.). BLUE-WINGED TEAL.

Hab. Eastern North America, northward to Labrador.

An example shot in Dumfriesshire in 1858 (wrongly recorded by Gray as killed in January, 1863) is now in Edinburgh Museum.

223. Querquedula crecca (Linn.). COMMON TEAL.

Hab. Palæarctic region; also Alaska and Greenland.

Male: chin black; head, cheeks and neck chestnut, with a patch of glossy dark green, narrowly bordered with buff, upon the ear-coverts and eye; mantle and flanks greyish-white, with fine blackish vermiculations; speculum metallic-green shading into purple, bordered above with black and below with white; rump dark brown; breast buff, marked with round black spots; centre of belly white; under tail-coverts black, bordered each side with buff; bill blackish; iris hazel; feet slate-brown. Length 14·25. Female: feathers of body chiefly dark brown in centre, with pale margins; speculum green and black; green patch on side of head absent.

Most common in winter, but breeds in small numbers in all parts of British Isles. Nest: near water in heather, rushes, rank grass, etc., or in morasses, bogs and damp places generally. Eggs: 8 or 9; cream-colour, tinged with green; 1·75 by 1·30.

224. Querquedula carolinensis (Gmel.). American Green-winged Teal.

Hab. North America. In winter migratory.

Has occurred in Hants (prior to 1840), near Scarborough (1851), and in South Devon (1879).

Genus CXXI. SPATULA, *Boie (1822).*

Bill considerably longer than head, narrow at base, much widened near tip; laminæ projecting.

225. Spatula clypeata (Linn.). Shoveller.

Hab. Palæarctic and Nearctic regions.

Male: head and upper neck dark green; lower neck, scapulars and tips of greater wing-coverts white, rest of coverts being light greyish-blue; speculum green; quills dusky; upper back brown; rump and tail blackish; belly chestnut; vent white; under tail-coverts black; bill bluish-slate; iris yellow; feet reddish-orange. Length nearly 20·00. Female: distinguished from the rather similar females of Gadwall and Wild Duck by its spoon-shaped bill. Most abundant as a winter visitor, but it breeds sparingly along the east side of England, from Norfolk northwards, and in many parts of Scotland; also in nearly every county of Ireland.

Nest: near water, but in tolerably dry spots; lined with down after eggs are laid as in the case of other ducks. Eggs: 8 to 10; similar to those of *A. boscas*, but smaller; 2·05 by 1·50.

GENUS CXXII. FULIGULA, *Stephens (1824).*

Bill nearly as long as head, somewhat elevated at base, depressed in middle, but rather prominent at tip; nail decurved; upper mandible slightly wider than lower, both laminated. Hind toe widely lobed.

226. Fuligula rufina (Pall.). RED-CRESTED POCHARD.

Hab. Southern Palæarctic region.

At long intervals about seventeen examples have been taken in England, one in Scotland, and one in Ireland.

227. Fuligula fuligula (Linn.). TUFTED DUCK.

Hab. Northern Palæarctic region.

Male: head, conspicuous dependent crest, neck, upper parts and breast black, glossed with violet on head and neck; *speculum white*; belly and sides pure white; under tail-coverts, black; bill bluish-slate; nail black; iris rich yellow; feet dark slate-blue, webs and claws blackish. Length 15·00. Female: all black parts of male, blackish-brown; belly and sides greyish; forehead mottled with white.

Breeds commonly in many localities in Nottinghamshire; also less frequently in various other counties in the east, north, and even south of England; while it now nests regularly in several parts of Scotland and Ireland; more abundant in most districts, however, during winter. Nest is hidden in sedges, rushes or long rank grass. Eggs: 8 to 12; pale greenish-buff; 2·30 by 1·55.

228. Fuligula marila (Linn.). SCAUP.

Hab. Northern Palæarctic and Nearctic regions.

Male: larger than *F. fuligula*, but resembling it in detail with the exception that upper back is white, each feather marked with numerous fine transverse lines of blackish-slate, and that head is not crested; colours of bill and feet similar. Length 17·50. Female: forehead and chin white; black upper parts of male dusky brown; upper back transversely marks on a brownish (instead of white) ground.

A regular and common winter visitor.

GENUS CXXIII. NYROCA, *Fleming (1822).*

Bill somewhat short and narrower than in *Fuligula* and of equal width throughout.

229. Nyroca ferina (Linn.). POCHARD.

Hab. Palæarctic region, except extreme north and east.

Male: head, throat and neck deep rufous; breast and a collar around lower neck black; mantle, wing-coverts, and secondaries greyish-white, finely vermiculated with transverse dusky lines; primaries blackish-slate on outer webs and tips, pale ash-grey on inner webs; rump and both upper and under tail-coverts black; below greyish-white, slightly vermiculated on sides; bill black at base and tip, bluish-slate in middle; iris orange-red; tarsi and toes bluish-slate, webs black. Length 18·50. Female: black parts of male dark brown; rest of plumage duller; upper throat whitish.

A common winter visitor but rare as a breeding species. It has nested in various English counties, also in several parts of Scotland and in a number of localities in the north and west of Ireland. Eggs: 7 or 8; light buffish-green; 2·35 by 1·60.

230. Nyroca nyroca (Güld.). White-eyed Duck.

Hab. Southern Palæarctic region.

An irregular straggler to eastern coasts of England, chiefly in winter and early spring; has also occurred in the south and west and two or three times near Edinburgh. In Ireland it has been taken five times. This is the *Fuligula ferruginea* of many authors.

Genus CXXIV. CLANGULA, *Leach (1819).*

Bill shorter than head, higher than wide at base, depressed towards tip; nail decurved; edges of upper mandible not inflected: concealing the laminæ. Head crested. Hind toe widely lobed.

231. Clangula clangula (Linn.). Golden-eye.

Hab. Northern Palæarctic region.

Male: head, throat and upper neck dark metallic green, with a patch of white at base of bill each side and a white collar on lower neck; back, lesser wing-coverts, bastard wing and inner secondaries black; outer secondaries, scapulars and greater coverts white, the two latter edged with black; breast and belly white; under wing-coverts black; thighs and vent dusky; bill bluish-black; iris orange-yellow; feet orange, webs dusky. Length 18·00.

Female: head, throat and upper neck dark brown, with a white collar around lower neck; crest and white facial patch absent; back, tail and scapulars dusky-slate; wing as in male; under parts chiefly white, with dusky mottlings on upper breast and sides; bill brownish-black, with an orange spot or bar near the nail.

A common winter visitor.

232. C. albeola (Linn.). BUFFEL-HEADED DUCK.

Hab. Northern Nearctic region.

Has occurred thrice in England and twice in Scotland.

GENUS CXXV. COSMONETTA, *Kaup (1829)*.

233. C. histrionica (Linn.). HARLEQUIN DUCK.

Hab. Iceland, Northern Nearctic region and N.E. Asia.

A very rare straggler. Mr. Whitaker has one obtained on the Yorkshire coast (1862), while Mr. R. W. Chase and the Rev. Julian G. Tuck have each an example obtained at the Farne Islands (1886). It also appears to have occurred twice in Scotland, but many years ago.

GENUS CXXVI. HARELDA, *Stephens (1824)*.

Bill shorter than head and narrowing towards tip, nail considerably decurved; laminæ partly exposed. Tail of 14 feathers, short, except two central ones which are much elongated and tapering.

234. Harelda glacialis (Linn.). LONG-TAILED DUCK.

Hab. Circumpolar region, northward of lat. 60°.

Male: head, throat and neck white; sides of head grey, followed by an oval patch of blackish-brown on each side of upper neck; breast, upper parts, wings and long central tail-feathers black; outer tail-feathers and long scapulars white ; belly white ; bill orange-pink, basal part and nail blackish ; iris reddish-brown ; feet slate-blue, webs blackish. Length 22·00. Female : crown blackish-brown ; hind neck, throat and upper breast grey ; side of head pale brown followed by a dark brown patch below ear-coverts, above which is a whitish stripe ; sides of lower neck whitish ; upper feathers dusky, with paler rufous margins : belly white ; central tail-feathers not lengthened.

A rather scarce winter visitor ; has long been supposed to breed in the Shetlands, but actual evidence is still lacking.

Genus CXXVII. HENICONETTA, *Gray (1840).*

235. Heniconetta stelleri (Pall.). Steller's Eider.

Hab. Northern Palæarctic and N.W. Nearctic regions.

A very rare straggler ; one has been killed in Norfolk (1830), and a second on the Yorkshire coast (1845).

Genus CXXVIII. SOMATERIA, *Boie (1822).*

Bill shorter than head and somewhat tapering, more or less swollen at base and extending upon forehead, where it is divided by a projecting point of short stiff feathers.

236. Somateria mollissima (Linn.). Eider Duck.

Hab. Northern portion of Western Palæarctic region.

Male : forehead, lores, crown, greater wing-coverts, primaries, rump and tail black ; rest of upper plumage and throat white, excepting nape, centre of occiput and a patch below ear-coverts, each side, which are pale green ; upper breast pinkish-buff ; under parts black with a white patch each side of vent ; bill greyish-green ; nail whitish ; iris brown ; feet pale green. Length 23·00. Female : feathers above light brown with dusky centres ; greater wing-coverts tipped with dull white, and secondaries margined with same; below paler and more distinctly barred. In summer male has white parts of plumage heavily mottled with black.

Breeds locally from Farne Islands northward to Orkneys, also in the Shetlands and Hebrides. Common on east side of England during winter. In Ireland only fifteen examples have been taken. Eggs: 5 or 6; pale greyish-green; 3·05 by 2·00.

237. Somateria spectabilis (Linn.). KING EIDER.

Hab. Circumpolar regions. In winter southward.

Seven or eight have occurred at the Farne Islands, one in Yorkshire, and two recently in Norfolk; it has been reported at irregular intervals from the Orkneys and Shetlands and several parts of Scotland, while in Ireland six have been obtained.

GENUS CXXIX. ŒDEMIA, *Fleming (1822).*

Bill large, much swollen at base of upper mandible and depressed towards tip; laminæ considerably developed.

238. Œdemia nigra (Linn.). COMMON SCOTER.

Hab. Northern Palæarctic region.

Male: black, glossed with green and purple above; bill black, with a yellow patch down centre of culmen; iris dark brown; feet brownish-black, webs black. Length 19·50. Female: duller and with a brown tinge; chin dull white; sides of head greyish-brown; bill with "knob" nearly absent. Length 18·50. Young: like female, but with throat whitish and under parts mottled with same.

Breeds in many localities in the north of the Scottish mainland, and is a common winter visitor to the rest of our coasts. Eggs: 6 to 8; buffish-white or light stone-colour; 2·50 by 1·75.

239. Œdemia fusca (Linn.). VELVET SCOTER.

Hab. Northern Palæarctic region.

Male: velvety-black; greater wing-coverts conspicuously tipped with white; below eye a small white patch; eyelids and iris white; bill light orange-yellow, knob and edges black, also a line from above each nostril to nail; feet orange-red; webs blackish. Length 22·00. Female: blackish-brown; white wing-bar obscure; a dull white spot on lores, and a larger one on ear-coverts; breast slightly mottled with white; bill slate-brown; iris brown.

A winter visitor in small numbers. Mr. R. Warren observed a pair in breeding plumage in Co. Sligo, June 24th, 1889.

240. Œdemia perspicillata (Linn.). SURF-SCOTER.

Hab. Northern Nearctic region.

Six or seven examples have been shot in the Orkneys, and others seen; has also been recorded from the Shetlands, Hebrides, Firth of Forth, and Aberdeenshire; while seven have been taken on the English and four on the Irish coasts.

GENUS CXXX. MERGUS, *Linnæus (1766).*

Bill shorter than head, straight and slender, except at base, which is tolerably stout; nail decurved, hooked; upper mandible only laminated, but cutting edges of both resolved into tooth-like serrations. Feathers of crown forming an elongated pendant crest.

241. Mergus albellus, Linn. SMEW.

Hab. Northern Palæarctic region.

Male: head, neck and throat white, with a triangular black patch on nape, and also one on the lores; back black; wings black and white; under parts white with two blackish crescentic lines on each side of breast; nail whitish, rest of bill and feet bluish-slate; iris ash-white. Length 17·00. Female: head and nape rufous-brown; mantle slate; other plumage duller than in male.

A winter visitor in small numbers.

Genus CXXXI. LOPHODYTES, *Reichenbach (1852).*

242. Lophodytes cucullatus (Linn.). Hooded Merganser.

Hab. North America. Partially migratory in winter.

Five examples in all have been obtained in Ireland and one in North Wales, while others are *said* to have been obtained in England and Scotland.

Genus CXXXII. MERGANSER, *Brisson (1760).*

Much as in *Mergus*, excepting that bill is longer than head.

243. Merganser merganser (Linn.). Goosander.

Hab. Northern Palæarctic region.

Male: head, throat and upper neck black, glossed with green; lower neck white; wing-coverts and outer secondaries white; rest of wing, scapulars and mantle black; rump and tail dark slate; under parts white, tinged with salmon-colour; bill deep red, nail black; iris reddish-brown; feet vermilion. Length 26·00. Female: chin whitish; head and upper neck reddish-brown; upper parts slate-grey, with darker shaft-streaks; greater coverts noticeably tipped with white, and outer secondaries also partly white; under parts whitish. Length 23·50.

A not uncommon winter visitor; also breeds in various parts of the Highlands, laying its eggs in a hollow tree upon a quantity of down. Eggs: 8 to 12; pale buff; 2·70 by 1·85.

244. M. serrator (Linn.). RED-BREASTED MERGANSER.

Hab. Northern Palæarctic and Nearctic regions.

Male: head, throat, upper neck, and a line down back of lower neck, black, glossed with green; rest of middle neck white; greater wing-coverts, secondaries and outer scapulars chiefly white; rest of wing and mantle black: rump and upper tail-coverts slate with black vermiculations; upper breast pale chestnut streaked with black, a patch of feathers on each side white broadly edged with black; under parts white; bill red, nail black; iris bright red; feet orange-red. Length 22·00. Female: smaller than female of *M. merganser*; mantle is dusky-brown, and greater wing-coverts are tipped with black, forming a bar across white wing-patch.

Breeds in the Highlands, Hebrides, Orkneys, Shetlands and in nearly all parts of Ireland. To England a winter visitor. Nest: well hidden in heather or long grass, under a boulder, or in a hole in the ground. Eggs; 7 to 10; pale buff, tinged with green; 2·55 by 1·75.

ORDER COLUMBÆ.

Family Columbidæ.

GENUS CXXXIII. COLUMBA, *Linnæus (1766).*

Bill moderate, compressed, tip slightly decurved; base furnished with a soft membrane in which nostrils are situated. Wings long, ample. Tarsus short, scaled in front; toes long, three before, one behind.

245. Columba palumbus (Linn.). RING-DOVE.

Hab. Western Palæarctic region.

Adult: plumage bluish-slate, paler on rump and browner on mantle; neck glossed with green and purple and having on each side a patch of white feathers; breast suffused with purplish-red; wing coverts noticeably margined with white. Length 16·00. Young: white feathers of neck absent.

Common everywhere; in winter gregarious. Nest: usually in trees and tall hedges; a slight platform of small sticks. Eggs: 2; pure white; shell oval and glossy; 1·60 by 1·20.

246. Columba œnas (Linn.). STOCK-DOVE.

Hab. Western Plæarctic region.

Differs from *C. palumbus* in lacking patch of white on each side of neck, although the feathers there are richly glossed with metallic green; there is less purplish-red on breast and no white on wings. Length 13·50. Young: metallic neck feathers absent in first plumage.

Locally distributed as far as South Scotland. In Ireland breeds down the eastern side. Eggs: laid in crowns of pollard trees, hollows in trees or cliffs, or even rabbit-burrows; creamy-white; 1·50 by 1·10.

247. Columba livia (Gmel.). ROCK-DOVE.

Hab. Western Palæarctic region.

This dove, the progenitor of our domestic pigeons, may be distinguished from the Stock-Dove by the *white rump*, the two conspicuous black bands crossing wing-coverts and secondaries respectively, and the white (instead of grey) under wing-coverts.

Found breeding chiefly in caves along the western side of England and all round the coasts of Scotland and Ireland; also at Flamborough Head on east coast of England. Eggs: whiter than those of *C. œnas*; 1·50 by 1·15.

Genus CXXXIV. TURTUR, *Selby (1835).*

Bill much as in *Columba* but more slender. Tail longer and somewhat graduated.

248. Turtur turtur (Linn.). Turtle-Dove.

Hab. Western Palæarctic region.

Adult: mantle rufous-brown; rest of upper parts bluish-grey, with a patch of black and white feathers on each side of neck; tail tipped with white; throat and breast light vinous-red; belly whitish. Length 11·50. Young: browner, and black and white feathers are absent from neck at first.

Common generally from May to September, although scarce in the extreme north and west. To Scotland chiefly a straggler. To Ireland a rare visitor; probably breeding occasionally in south. Frequents woods and copses, making a slighter nest than Ring-Dove's and lower down. Eggs: creamy-white; size 1·20 by ·90.

Genus CXXXV. ECTOPISTES, *Swainson (1827).*

249. E. migratorius (Linn.). Passenger Pigeon.

Hab. Eastern Nearctic region.

Five examples have been obtained, but it is by no means certain how many of these were wild.

ORDER PTEROCLETES.

Family Pteroclidæ.

Genus CXXXVI. SYRRHAPTES, *Illiger (1811).*

Bill small, upper mandible slightly decurved from base; nostrils basal, concealed. Wings very long, pointed. Two middle tail-feathers very long and tapering. Tarsi very short, feathered; toes, three only, directed forward and united by a membrane.

250. Syrrhaptes paradoxus (Pall.). PALLAS'S SAND-GROUSE.

Hab. Central Asia. In autumn migrating southward, eastward or westward, irregularly invading Europe at this season, or on return migration in spring.

The first irruption occurred in 1859; a second in 1863-64; a third and fourth in 1872 and 1876; a fifth and greatest in 1888-89. The latter extended over whole of British Isles, and eggs were found in Norfolk, Nottinghamshire and Yorkshire, and a nestling in Scotland.

ORDER GALLINÆ.

Family Phasianidæ.

GENUS CXXXVII. PHASIANUS, *Linnæus (1766).*

Bill moderate, stout, strong, upper mandible decurved at tip. Tail very long, graduated. Tarsus furnished behind with a sharp spur in male; toes three in front, united to first joint, one behind.

251. Phasianus colchicus, Linn. PHEASANT.

Hab. S.E. Europe and Asia Minor (originally); also most of Europe (introduced).

Common in preserves; the offspring of hybrids between this and the Red-necked Pheasant appear, however, to be more prevalent than pure-bred birds.

GENUS CXXXVIII. CACCABIS, *Kaup (1829).*

252. Caccabis rufa (Linn.). RED-LEGGED PARTRIDGE.

Hab. Western Europe.

Introduced more than a century ago; now common in S.E. of England from Sussex to Norfolk.

GENUS CXXXIX. PERDIX, *Brisson (1760).*

Bill much as in *Phasianus*, but shorter. Tarsus lacking the spur. Tail rather short, slightly rounded.

253. Perdix cinerea, Lath. COMMON PARTRIDGE.

Hab. Temperate Europe and Western Asia.

Common throughout British Isles.

GENUS CXL. COTURNIX, *Bonnaterre (1790).*

Characters much as in *Perdix*.

254. Coturnix coturnix, Linn. QUAIL.

Hab. Palæarctic region and whole of Africa. In winter migrating from N. Europe.

Easily distinguished from the Partridge by its small size; length 7·00; wing 4·30. Male differs from female in having a blackish patch on upper throat.

Locally but widely distributed in summer, while a few often remain through the winter. Eggs: 7 to 10; creamy-white, variably clouded or blotched with chocolate-brown; 1·10 by ·90.

Family Tetraonidæ.

GENUS CXLI. LAGOPUS, *Brisson (1760).*

Bill short, strong, the base clothed with feathers, tip of upper mandible decurved. Tail moderate, even. Feet feathered to claws. Hind toe short.

255. Lagopus mutus (Montin). PTARMIGAN.

Hab. Mountains of Europe and Central Asia.

Male: bare skin over eye red; lores and a stripe through eye black; crown and breast mottled with black; upper parts freckled and barred with grey and brown; wing-quills white; tail black, tipped with white; belly white. Length 14·50. Female: plumage tawny, with blackish bars; black of lores almost absent, but wing-quills are white. In winter both sexes are white, but male still shows the black lores.

Resident in the mountains of the Northern Highlands and the Hebrides.

256. Lagopus scoticus (Lath.). RED GROUSE.

Hab. British Islands.

Common on the moorlands and mountains of Scotland, Northern England and Wales; also in many parts of Ireland. Eggs: 7 to 10; yellowish-white, blotched and marbled closely with rich reddish-brown; 1·75 by 1·25.

GENUS CXLII. TETRAO, *Linnæus (1766).*

Characters much as in *Lagopus*, but tail is longer and consists of 18 instead of 16 feathers. Tarsus feathered but toes bare.

257. Tetrao tetrix, Linn. BLACK GROUSE.

Hab. Northern Palæarctic region.

Male: bluish-black, with a white band across wing; under tail-coverts also white; tail forked, the feathers curved outwards. Length 21·00. Female (Grey Hen): plumage pale reddish-brown, barred all over with blackish-brown; also smaller, and tail not forked.

Distributed over Great Britain, but confined to the moorlands and mountains. It is not found in Ireland. Eggs: 7 to 10; pale yellowish, spotted with dark orange-brown; 2·00 by 1·40.

258. Tetrao urogallus, Linn. CAPERCAILLIE.

Hab. Northern Palæarctic region.

This fine bird became extinct in both Ireland and

Scotland a little more than a century ago, but it has since been re-introduced into the Highlands where it is now rapidly spreading.

ORDER HEMIPODII.

Family Turnicidæ.

GENUS CXLIII. TURNIX, *Bonnaterre (1790)*

259. Turnix sylvatica, Desfont. ANDALUSIAN HEMIPODE.

Hab. Southern Spain, Sicily and N. Africa.

Two examples have been taken in Oxfordshire (1844), and a third near Huddersfield (1865).

ORDER FULICARIÆ.

Family Rallidæ.

GENUS CXLIV. RALLUS, *Linnæus (1766).*

Bill longer than head, slender, nearly cylindrical towards tip; very slightly decurved; upper mandible laterally grooved. Tarsus rather long; toes long, slender, three in front, one behind.

260. Rallus aquaticus, Linn. WATER-RAIL.

Hab. Western Palæarctic region.

Adult: feathers above olive-brown with black centres; cheeks, throat and breast dull slate: sides blackish barred with white; under tail-coverts buffish-white; bill deep red; iris hazel; tarsi pale brown. Length 10·25. Young: under parts whitish, spotted and barred with brown. Nestling: jet black with a whitish bill.

Generally distributed and resident. Nest: usually placed in a clump of sedges or reeds and composed of aquatic herbage. Eggs: 7 or 8; buffish-white, sparingly spotted with grey and light reddish-brown; 1·40 by 1·05.

GENUS CXLV. PORZANA, *Vieillot (1816).*

Bill shorter than head, somewhat compressed, higher than wide at base, upper mandible slightly decurved at tip.

261. Porzana porzana (Linn.). SPOTTED CRAKE.

Hab. Temperate Europe and W. Asia. In winter southward to Africa and India.

Adult: feathers above olive-brown with darker centres and sprinkled all over with small white spots; cap dark brown; sides of head, throat and belly greyish; breast light brown spotted with white; flanks dusky barred with white; bill yellow, reddish at base; tarsi yellowish-green. Length 9·00.

A very local summer visitor to England, and irregularly to Wales and South Scotland. A rare straggler to Ireland, although apparently having bred on two occasions.

262. Porzana bailloni (Vieill.). BAILLON'S CRAKE.

Hab. Southern Palæarctic region and all Africa.

A very rare visitor, chiefly to the eastern counties, where it is said to have bred in 1858 and again in 1866. It has occurred twice in Scotland and twice in Ireland.

263. Porzana parva (Scop.). LITTLE CRAKE.

Hab. Temperate Europe and S.W. Asia.

Scarcely a more frequent visitor than the last. It has been recorded thirteen or fourteen times from Norfolk, also from every other southern and eastern county, but more rarely from the west, and only once each from Scotland and Ireland.

GENUS CXLVI. CREX, *Bechstein (1803).*

Differs not greatly from *Porzana*; bill is rather stouter, and upper mandible is gently decurved from forehead.

264. Crex crex (Linn.). LAND-RAIL.

Hab. Western Palæarctic region.

Adult: feathers above yellowish-buff with blackish centres; wings chestnut brown; sides of head greyish; throat white; breast tinged with brown; flanks buff, barred with dark brown; bill and tarsi pale brown. Length 10·50.

Common from May to September in meadows and pastures and also cornfields. The familiar and persistent cry is commonly heard through the night as well as by day. Nest: on ground, usually among long meadow-grass. Eggs: 7 to 9; reddish-white, sparingly spotted with dull red and grey; 1·45 by 1.10.

GENUS CXLVII. GALLINULA, *Brisson (1760).*

Bill about as long as head, much compressed; gonys ascending, culmen decurved; base of upper mandible extended upon forehead, forming a frontal plate. Toes margined throughout their length with a membrane.

265. Gallinula chloropus (Linn.). MOOR-HEN.

Hab. Europe, Asia and Africa.

Adult: above dusky-brown tinged with olive; head, nape and under parts blackish-slate; flanks striped with white; under tail-coverts pure white; frontal plate and base of bill red, tip yellow. Length 12·50. Young: bill and frontal plate greenish; throat whitish.

Familiar and abundant on ponds, rivers and ditches.

GENUS CXLVIII. FULICA, *Brisson (1760).*

Much like *Gallinula*, but membrane bordering toes is widened and scalloped, forming a series of lobes.

266. Fulica atra, Linn. COOT.

Hab. Europe and temperate Asia.

Adult: above dusky-slate, with a white bar across wing; below dull black; frontal plate white; bill pinkish-white; iris deep red; feet dark green. Length 16·50. Young: duller; throat whitish.

A fairly common resident, frequenting lakes, rivers and large ponds. Nest is a mass of sedges, flags, etc., supported by a bunch of reeds or rushes growing in the water. Eggs: 7 or 8; pale buff, sparingly marked with minute spots of dark brown; 1·95 by 1·50.

ORDER ALECTORIDES.

Family Gruidæ.

GENUS CXLIX. GRUS, *Bechstein (1793).*

267. Grus grus (Linn.). CRANE.

Hab. Palæarctic region. In winter southward to Africa and India.

Chiefly occurring now as a straggler on migration in the Orkneys and Shetlands, but very rarely found in England. About six have been shot in Ireland.

Family Otididæ.

GENUS CL. OTIS, *Linnæus (1766).*

268. Otis tarda, Linn. GREAT BUSTARD.

Hab. Southern Palæarctic region.

As an indigenous species this fine bird became extinct in England rather more than fifty years ago. It is now known as a very irregular straggler in winter.

269. Otis tetrax, Linn. LITTLE BUSTARD.

Hab. Temperate Europe, N. Africa and S.W. Asia.

An irregular winter visitor; has occurred on nearly fifty occasions in England, but only four or five times in Scotland and six times in Ireland.

GENUS CLI. HOUBARA, *Bonaparte (1831).*

270. Houbara macqueenii (Gray). MACQUEEN'S BUSTARD.

Hab. Central Asia. In winter reaching India.

An example was shot in Lincolnshire (1847), and a second in Yorkshire (1892).

ORDER LIMICOLÆ.

Family Œdicnemidæ.

GENUS CLII. ŒDICNEMUS, *Temminck (1815).*

Bill shorter than head, straight, strong; base of upper mandible depressed; culmen arched, slightly decurved at tip; gonys with an abrupt angle and ascending to tip. Nostrils lateral, nearly in middle of upper mandible. Tarsus rather long; hind toe absent.

271. Œ. œdicnemus (Linn.). STONE-CURLEW.

Hab. Palæarctic region. In winter partly migratory.

Adult: feathers above sandy-brown with dusky centres; greater wing-coverts tipped with white, and median-coverts sub-terminally banded with the same; quills blackish; a short streak above eye, a longer one below, and also the throat, buffish-white; upper breast and sides buff with blackish streaks; belly white; base of bill greenish-yellow, rest black; iris very large, rich yellow; tarsi yellow; Length 16·00.

Found from April to October on downs, wolds and open heaths of south and east of England, while a few remain through the winter. In Scotland one or two have been taken, and also six or eight in Ireland. Eggs: 2; pale yellowish-buff, streaked with grey and blotched with deep brown; 2·15 by 1·55; laid on ground in a slight depression or among stones.

Family Glareolidæ.

GENUS CLIII. GLAREOLA, *Brisson (1760).*

272. Glareola pratincola (Linn.). PRATINCOLE.

Hab. S. Europe, W. Asia and N. Africa.

An irregular visitor on migration to England; one has also been taken in the Shetlands, and probably one in Ireland.

Family Charadriidæ.

GENUS CLIV. CURSORIUS, *Latham (1790).*

273. C. gallicus (Gmel.). CREAM-COLOURED COURSER.

Hab. North Africa and S.W. Asia.

A rare straggler to England, about twenty examples having been obtained. Has also occurred once in Scotland (1868).

Genus CLV. CHARADRIUS, *Linnæus (1766).*

Bill rather shorter than head, much as in *Œdicnemus*, but more slender. Wings long and pointed, 1st primary longest. Tarsus moderately long, reticulated; three toes in front united by a membrane at base; hind toe obsolete.

274. Charadrius pluvialis, Linn. Golden Plover.

Hab. Northern Europe and Eastern Siberia. In winter reaching South Africa.

Adult: above black, closely barred and spotted with golden-yellow and greyish-white; forehead and a line over eye white; cheeks and under parts black bordered with white on sides of lower breast and belly; *axillaries pure white;* iris dark brown; bill and tarsi slate-black. Length 10·25. In winter yellower above and white below, with a dusky zone on breast.

Common during migration and in winter, frequenting the coasts and low-lying districts. A fair number breed in the south-west and north of England, in Wales, Scotland and the mountainous parts of Ireland. Eggs: 4; greyish-yellow to warm buff, heavily blotched and spotted with blackish-brown; 1·95 by 1·40.

275. Charadrius dominicus (Müller). American Golden Plover.

Hab. Arctic America from Alaska to Greenland.

An example of this smaller species was purchased by Mr. J. H. Gurney in Leadenhall Market (1882) while a second was taken in Perthshire (1883).

275a. Ch. dominicus fulvus (Gmel.). Pacific Golden Plover.

Hab. N.E. Asia.

One was obtained in Leadenhall Market (1874) and a second shot in the Orkneys (1887).

GENUS CLVI. SQUATAROLA, *Leach (1816).*

Much as in *Charadrius* but feet with ***hind toe present*** although small and elevated.

276. Squatarola helvetica (Linn.). GREY PLOVER.

Hab. Circumpolar regions. In winter migrating southward over greater part of globe.

Adult: above dull black, barred and mottled with greyish-white; forehead and a broad stripe over eye white; cheeks, throat, breast and upper belly black; under wing-coverts, vent and under tail-coverts white; *axillaries black*; bill and feet slate-black. Length 11·00. In winter black of face, throat and breast is lost, but latter shows a few dusky mottlings.

A common visitor in autumn or winter.

GENUS CLVII. ÆGIALITIS, *Boie (1822).*

Bill rather shorter and more slender than in *Charadrius.* Feet similar; three toes only; tarsus slender.

277. Ægialitis hiaticula (Linn.). RINGED PLOVER.

Hab. Whole Palæarctic region.

Adult: above dull brown, with some white at base of wing-quills; outermost tail-feather each side pure white, rest with decreasing white tips; forehead and a streak above ear-coverts white; crown, ear-coverts and space below eye black; chin, upper throat and sides of neck white, bordering a broad black pectoral band, below which under parts are white; bill orange at base, rest black; iris hazel; tarsi orange. Length 7·50. Description is that of British individuals, which have been separated as *Æ. hiaticula major* by Seebohm, the birds breeding in Temperate Europe, Central Asia and N. Africa being smaller and brighter, with a wing measurement of 5·00 as against 5·30 to 5·40 in our birds. This small form often visits our south coast in spring.

The larger form is common on the flat portions of our coasts and more or less resident. Eggs: 4; pear-shaped; pale buff, evenly marked with small spots and streaks of black; 1·40 by 1·00.

278. Ægialitis dubia (Scop.). LITTLE RINGED PLOVER.

Hab. Palæarctic region. In winter migrating to Africa.

A rare wanderer in spring or autumn to England, only six genuine examples having been obtained.

279. Ægialitis alexandrina (L.). KENTISH PLOVER.

Hab. Temperate Palæarctic region.

Male: distinguished from *Æ. hiaticula* by the *incomplete pectoral band*, there being a patch of black on each side only of upper breast; bill, legs and feet also are *black*. Length 6·75. Female: black patches on sides of breast dark brown; black of fore-crown absent.

Found locally from April to September on eastern and southern coasts of England; it rarely strays farther north than Yorkshire or to west side, although one has been shot in Ireland. Eggs: 3; more heavily marked than those of Ringed Plover and smaller; 1·25 by ·90.

280. Ægialitis vocifera (Linn.). KILDEER PLOVER.

Hab. Temperate North America. In winter migratory.

One is said to have been shot in Hampshire (1857); a second has since been taken in the Scilly Isles (1885).

GENUS CLVIII. EUDROMIAS, *Brehm (1831).*

Possessing much of characteristics of *Ægialitis* and *Charadrius*, but tarsi scaled in front instead of reticulated.

281. Eudromias morinellus (Linn.). DOTTEREL.

Hab. Northern and Central Europe and W. Asia. In winter reaching N. Africa.

Adult: feathers above greyish-brown, with paler or sandy margins; crown nearly black; above eye a broad white stripe, running downward and backward to hind neck; forehead, lores and cheeks white with small dusky spots; upper throat white; lower throat greyish-brown, becoming black on margin of a narrow white pectoral band; lower breast and sides rufous; centre of abdomen black; vent white; bill blackish; tarsi dark greenish-slate. Length 8·75. In winter black of belly is absent.

Probably still breeds in the Lake district and in several parts of Scotland, but elsewhere is only met with on migration. A very rare visitor to Ireland.

GENUS CLIX. VANELLUS, *Brisson (1760).*

Bill as in *Charadrius.* Wings broad and rounded. Tarsus scutellated anteriorly; hind toe present but small. Head with an erectile crest of long pointed recurved feathers.

282. Vanellus vanellus (Linn.). LAPWING.

Hab. Palæarctic region.

Breeds commonly almost everywhere; in winter wholly gregarious and partially migratory. Frequents waste land, dry marshes, or meadows. Eggs: 4; varying from warm buff to light olive, blotched and spotted with brownish-black; 1·75 by 1·30. Familiar note is expressed in its local names, "Peewit," "Peesweep," etc., but in spring the male utters more varied love-notes. Also known as "Green Plover."

GENUS CLX. CHÆTUSIA, *Bonaparte (1841)*.

283. Chætusia gregaria (Pall.). SOCIABLE LAPWING.

Hab. S.E. Russia and S.W. Asia. Wintering in India and N.E. Africa.

An example shot in Lancashire about 1860 was at first recorded as *Cursorius gallicus*, but afterwards found to belong to this species.

GENUS CLXI. STREPSILAS, *Illiger (1811)*.

Bill as long as head, rather stout at base, tapering to a blunt point, upper mandible slightly longer than lower. Hind toe very small.

284. Strepsilas interpres (Linn.). TURNSTONE.

Hab. Circumpolar regions.

Male: head, neck and breast black, variegated with white; mantle and wings black with chestnut margins to the feathers; rump and base of tail-feathers white; upper throat and belly also white; bill black; iris hazel; tarsi orange-red, claws black. Length 8·50. Female duller.

Common on our coasts during migration, while a proportion remain through the winter.

GENUS CLXII. HÆMATOPUS, *Linnæus (1766)*.

Bill much longer than head, straight, mandibles nearly equal, strong, much compressed towards tip which is abruptly truncated. Tarsus rather long, stout; hind toe absent.

285. Hæmatopus ostralegus, L. OYSTER-CATCHER.

Hab. Northern Europe and Central Asia.

Adult: chin, throat and upper parts black, except greater wing-coverts, part of secondaries, and the rump and upper tail-coverts, which are white; below white; bill red, yellowish at base and tip; iris deep red; feet purplish-pink. Length 16·00. Young: browner above.

Breeds on nearly all the flat shores of the British Isles; in winter occurring in flocks and partly migratory. Eggs: usually 3; pale buffish-yellow, rather sparingly spotted, scrolled or even blotched with brownish-black; 2.25 by 1·60.

Family Scolopacidæ.

Genus CLXIII. RECURVIROSTRA, *Linnæus (1766).*

Bill very long and slender, flexible, flattened and curving upwards. Tibia chiefly naked; tarsus long; three front toes joined by a web as far as second joint; hind toe small and elevated.

286. Recurvirostra avocetta, Linn. AVOCET.

Hab. Temperate Europe and Asia; also Africa.

Now a rare visitor in spring to the south and east of England; not known to have bred for many years; few stragglers reach Scotland, while in Ireland only three have been taken (one in autumn).

Genus CLXIV. HIMANTOPUS, *Brisson (1760).*

287. H. himantopus (Linn.). BLACK-WINGED STILT.

Hab. Southern Palæarctic and Oriental regions.

A rare and irregular spring visitor, chiefly to the east and south of England, but it has occurred as far north as the Shetlands, and five or six times in Ireland.

Genus CLXV. PHALAROPUS, *Brisson (1760).*

Bill longer than head, straight, somewhat slender and tapering, base depressed. Anterior toes furnished with broad membraneous lobes; hind toe present, but moderate.

288. Phalaropus hyperboreus (Linn.). RED-NECKED PHALAROPE.

Hab. Northern Palæarctic and Nearctic regions.

Adult: above slate-grey, turning to blackish on rump and wings; scapulars and some of mantle-feathers margined with reddish-buff; greater coverts broadly tipped, and secondaries edged, with white; chin white; throat and sides of neck bright rufous; upper breast slate; under parts white; bill black; feet dark green. Length 7·00. In winter upper parts are duller with white margins to the feathers; forehead, cheeks and under parts pure white.

Breeds sparingly in the Orkneys, Shetlands and some of the Hebrides, migrating southward in winter, but not touching our coasts very frequently on passage; in Ireland, in fact, only one example has been obtained (1891).

GENUS CLXVI. CRYMOPHILUS, *Vieillot (1816).*

Bill very little longer than head, depressed at base, but thickening near tip, instead of tapering as in *Phalaropus*.

289. Crymophilus fulicarius (L.). GREY PHALAROPE.

Hab. Circumpolar regions. Wintering down to Equator.

An irregular winter visitor, sometimes, however, appearing in considerable numbers. Adult in winter has upper parts light bluish-grey; forehead and under parts white; occiput black; wings dusky, with secondaries and tips of wing-coverts white; bill blackish; feet yellowish. Length 7·50.

GENUS CLXVII. SCOLOPAX, *Brisson (1760).*

Bill very long, straight, slender; upper mandible slighly longer than lower, its tip bent over end of latter.

Nostrils basal, lateral. Tibia feathered; anterior toes without membranes; hind toe moderate.

290. Scolopax rusticola, Linn. WOODCOCK.

Hab. Palæarctic regions.

Adult: feathers of upper parts rufous, freckled and barred with black and tipped with grey; chin white; under parts greyish-white, suffused with buff and barred with brown; bill dusky-brown; iris dark brown; tarsi slate. Length 12·00.

Common everywhere in winter, but also nesting irregularly in most districts. Eggs: 4; pale buff, blotched with pale and dark reddish-brown, and with underlying lilac blotches; shape somewhat globular; 1·70 by 1·30; laid in a depression among dead leaves in thick coverts.

GENUS CLXVIII. GALLINAGO, *Leach (1816).*

Differs from *Scolopax* chiefly in having the inner secondaries about as long as the primaries, and the tibiæ bare on lower portion.

291. Gallinago major (Gmel.). GREAT SNIPE.

Hab. Northern Europe and Western Siberia. In winter southward to whole of Africa.

Occurs annually in east and south of England during autumn, but more rarely in west; not more than ten have been killed in Scotland, and only four in Ireland. It may be distinguished from *G. gallinago* by the wing-coverts being tipped with white, and the four outer tail-feathers on each side being almost entirely white; tail-feathers also are 16 in number, instead of 14. Length 8·75; wing 5·50.

292. Gallingo gallinago (Linn.). COMMON SNIPE.

Hab. Palæarctic region.

Adult: above variegated with blackish and pale buff; chin white; throat and breast pale brown, with dusky spots and mottlings; belly white. Length 8·00; wing 5·25. "Sabine's Snipe" is now admitted to be a melangism of this species.

Breeds in most parts of the British Isles, but is more common everywhere in winter. Eggs: pale yellowish, with an olive tinge, blotched with reddish-brown and blackish, and with underlying lilac marks; 1·60 by 1·15.

GENUS CLXIX. LYMNOCRYPTES, *Kaup (1829).*

Has two notches on each side of the posterior margin of the sternum, whereas in the last two genera there is but one. Tail-feathers also are but 12 in number.

293. Lymnocryptes gallinula (Linn.). JACK-SNIPE.

Hab. Northern Palæarctic region.

A tolerably common winter visitor. It is easily distinguished from the Common Snipe by its smaller size and fewer tail feathers. Length 6·25 to 6·50; wing 4·30.

GENUS CLXX. LIMICOLA, *Koch (1816).*

294. Limicola platyrhyncha (Temm.). BROAD-BILLED SANDPIPER.

Hab. North Europe. In winter moving southward. Has been obtained five times in Norfolk, twice in Sussex, once in Yorkshire, and once in Ireland.

GENUS CLXXI. TRINGA, *Linnæus (1766).*

Bill a little longer than head, straight or slightly decurved, slender, base compressed, dilated near point which is rather obtuse. Tarsus moderately long, slender; hind toe small, elevated; front toes unwebbed.

295. Tringa maculata, Vieillot. PECTORAL SANDPIPER.

Hab. Arctic America. In winter reaching S. America.

About two dozen examples have been shot in various parts of England, two or three of them in spring, also two in Scotland and one in Ireland.

296. Tringa acuminata (Horsf.). SHARP-TAILED SANDPIPER.

Hab. N.E. Siberia. In winter reaching Australia.

In 1892 an example was shot on Breydon Broad, Norfolk, and recorded by Mr. Southwell (Zool. 1892, p. 356) who has since stated (*tom. cit.* p. 405) that a supposed example of *T. maculata*, in Norwich Museum (a reputed Norfolk bird) also belongs to this species.

297. Tringa fuscicollis, Vieill. BONAPARTE'S SANDPIPER.

Hab. Arctic America. In winter reaching Equator.

More than a dozen examples have been obtained in England, and one in Ireland.

298. Tringa alpina, Linn. DUNLIN.

Hab. Palæarctic region.

Adult: feathers above rufous with black centres; on rump slate-brown with darker centres; wing dark brown, greater coverts tipped with white, secondaries with white basal spots and narrow white tips; throat and upper breast buffish-white streaked with black; lower breast black; rest of under parts white, under tail-coverts streaked with black; bill and tarsi black. Length 7·00; wing 4·25. In winter greyish-brown above; secondaries chiefly white;

greater coverts with increased white tips; outer feathers of upper tail-coverts white; black of lower breast absent; throat white; upper breast with a mottled dusky band.

Breeds in the moorlands and marshes of the north and south-west of England, throughout Scotland and several parts of Ireland. It is abundant on our coasts during winter. Eggs: 4; light greenish-grey or buffish-green blotched and spotted with reddish-brown and brownish-black; 1·40 by 1·00.

299. Tringa minuta, Leisl. LITTLE STINT.

Hab. Arctic regions of Europe and Asia.

Adult in winter resembles Dunlin, but is much smaller. Length 5·25; wing 3·75; bill ·75. It is a regular autumnal visitor to the coasts of England and N.E. Ireland, but less commonly to Scotland.

300. Tringa minutilla, Vieill. AMERICAN STINT.

Hab. Arctic America. In winter reaching Brazil.

One has been shot in Cornwall (1853), and two in Devon (1869, 1892).

301. Tringa temmincki, Leisl. TEMMINCK'S STINT.

Hab. Arctic regions of Europe and Asia.

An irregular visitor on migration to the east and south of England. Resembles *T. minuta*, and measurements are identical, but it has the two outer tail-quills on each side *white*; tarsi are greyish-olive in this species and black in *T. minuta*.

302. Tringa subarquata (Güld.). CURLEW SANDPIPER.

Hab. Polar regions (breeding-ground unknown).

Adult: above chestnut-brown with black centres and narrow greyish-white margins to all the feathers; rump duller and greyer; upper tail-coverts chiefly white; greater wing-coverts and secondaries edged and tipped with white; below light chestnut-red; sides and vent white; bill black, noticeably decurved; tarsi black. Length 7·50; wing 5·00. In winter chiefly greyish-brown above; upper tail-coverts white; below white, slightly speckled with dark brown on upper breast.

A fairly common visitor in spring and autumn.

303. Tringa striata, Linn. PURPLE SANDPIPER.

Hab. N. Europe and N.E. America.

Adult (winter): feathers above blackish, with greyish margins and with a slight purplish gloss; outside tail-coverts chiefly white; most of inner secondaries white; chin greyish-white; throat and upper breast slate-brown; under parts white, with dusky mottlings and streaks on sides. In spring mantle feathers are margined with reddish-brown; rump and upper tail-coverts black; throat and upper breast greyish-white, closely streaked with dark brown and sides spotted with same. Length 7·30; wing 5·00; bill straight, dark brown; tarsi yellowish.

Moderately common during winter on our coasts, often remaining until May.

304. Tringa canutus, Linn. KNOT.

Hab. Polar regions, north of Arctic America.

In both spring and autumn this species very much resembles the Curlew Sandpiper but may be instantly distinguished by its larger size and straight bill. Length 9·00; wing 6·40; bill and tarsi black; iris dark brown. A common visitor during migration and in winter.

GENUS CLXXII. PAVONCELLA, *Leach (1816).*

Bill rather longer than head, straight, much as in *Tringa.* Greater part of tibia naked; outer toe united to middle at base by a membrane. Neck of male furnished in spring with elongated curled feathers capable of dilation.

305. Pavoncella pugnax (Linn.). RUFF.

Hab. Northern Palæarctic region. In winter reaching Africa, India and (casually) N.E. America.

The sexes are alike except during the breeding season. Female (Reeve): feathers above and on throat and breast dark brown with buffish or whitish margins; rump darker; greater wing-coverts chiefly black, glossed with green and tipped with white; belly white; bill dark brown; tarsi yellowish-brown; length 10·00; wing 6·00. Male: length 12·00.

Still breeds casually in Lincolnshire and Norfolk, but is now chiefly known as a scarce migrational visitor.

GENUS CLXXIII. CALIDRIS, *Cuvier (1800).*

Bill about as long as head, straight, much as in *Tringa.* Tarsus rather short; *hind toe absent.*

306. Calidris arenaria (Linn.). SANDERLING.

Hab. Circumpolar regions.

Adult: above much as in Curlew Sandpiper; cheeks, throat and upper breast pale chestnut-red spotted with dusky brown; under parts white. In winter upper parts are chiefly light grey and under parts white. Length 6·90; wing 4·75; bill black; tarsi dusky olive.

Fairly common on our coasts during migration.

GENUS CLXXIV. TRYNGITES, *Cabanis (1856).*

307. Tryngites rufescens (Vieill.). BUFF-BREASTED SANDPIPER.

Hab. Arctic America and N.E. Asia.

More than a dozen examples have been obtained in England and three in Ireland; one is also said to have been taken in Scotland.

Genus CLXXV. BARTRAMIA, *Lesson (1831).*

308. Bartramia longicauda (Bech.). Bartram's Sandpiper.

Hab. North America (chiefly inland).

Has occurred eight times in England and twice in Ireland.

Genus CLXXVI. TRINGOIDES, *Bonaparte (1831).*

Bill scarcely longer than head, straight, slender. Tarsus moderate. Hind toe rather small; middle toe united to outer at base.

309. Tringoides hypoleucus (Linn.). Common Sandpiper.

Hab. Palæarctic region.

Adult: feathers above bronze-brown, with blackish central spots or bars; greater wing-coverts tipped with white; secondaries with white basal patches and tips; tail tipped with white, two outer feathers each side chiefly white; chin and line above eye white; throat and breast buffish grey with dusky streaks; below white; bill dusky, paler below; iris brown; tarsi greyish-green. Length 7·00; wing 4·20. In winter of a more uniform olive-brown above and less marked below.

Breeds rather sparingly in the south of England, but more commonly in west and north, and throughout Scotland and Ireland. Frequents the shingly margins of streams and lakes, arriving in April and leaving again in August. Eggs: 4; pale reddish-buff or rufous-white, spotted with reddish-brown; 1·40 by 1·00.

Genus CLXXVII. RHYACOPHILUS, *Kaup (1829).*

Bill a little longer than head. Tarsus rather long, exceeding the length of bill.

310. Rhyacophilus glareola (Gmel.). WOOD-SANDPIPER.

Hab. Palæarctic region. Wintering in Africa.

Distinguished from next species by buffish-white spots on upper parts being larger and more numerous; *axillaries white*, slightly speckled with brown; also smaller and with a longer tarsus. Length 7·70; wing 4·80; bill 1·10; tarsus 1·40.

A scarce visitor to England during migration. Rarely reaches Scotland; has once occurred in Ireland.

GENUS CLXXVIII. HELODROMAS, *Kaup (1829).*

Bill considerably longer than head. Tarsus moderate.

311. Helodromas ochropus (Linn.). GREEN SANDPIPER.

Hab. Northern Palæarctic region.

Adult: above olive-brown minutely spotted with white; upper tail-coverts pure white; outer tail-feathers entirely white, rest with blackish bars; throat and breast whitish with greyish-brown streaks; under parts white; axillaries blackish-brown with narrow and distinct bars of white; bill dusky brown, paler below; tarsi slate, tinged with green. Length 8·00; wing 5.40.

A common visitor during migration.

312. Helodromas solitarius (Wils.). SOLITARY SANDPIPER.

Hab. N. America. In winter reaching S. America,

One was killed many years ago on the Clyde, a second in the Scilly Isles (1882), and a third in Cornwall (1884).

GENUS CLXXIX. TOTANUS, *Bechstein (1803).*

Bill longer than head, straight, upper mandible slightly decurved at extremity. Tibia bare on lower half. Tarsus rather long. Outer toe united to middle at base.

313. Totanus calidris (Linn.). REDSHANK.

Hab. Palæarctic region. Wintering in Africa.

Adult: feathers above light brown, centred and edged with blackish-brown; secondaries chiefly white; rump white, slightly mottled with black; tail white, with blackish bars, central pair brown; cheeks and under parts white, throat and breast streaked and sides barred with dark brown; belly white in centre; bill blackish, reddish at base; tarsi orange-red. Length: male 9·50; wing 5·90; female 10·00; wing 6·20. In winter slate-brown above; rump and secondaries white; below white, sparingly marked.

Breeds in small numbers everywhere in marshy districts, but the majority leave us in winter. Eggs: yellowish-buff, marked with spots and small blotches of dark purplish-brown; 1·70 by 1·20. Alarm note is a loud *took* or *too-ik*, uttered as the bird hovers in the air; it also utters a kind of song in the spring.

314 Totanus fuscus (Linn.). SPOTTED REDSHANK.

Hab. Northern Palæarctic region.

Adult: above sooty-black; rump white; *secondaries white, barred irregularly with dusky-brown;* tail dusky-brown, with narrow white bars; upper and under tail-coverts barred with black and white; under parts sooty-black, mottled with white on belly; under wing-coverts and axillaries white as in *T. calidris.* In winter darker above than latter, and with white edges to most of the feathers; rump and secondaries as in spring; below white, tinged and streaked with dusky-grey on breast. Length 11·25; wing 6·50.

An irregular but not uncommon migrational visitor to East Anglia; very few have been taken in Scotland; in Ireland has occurred seven times.

315. Totanus flavipes (Gmel.). YELLOWSHANK.

Hab. Arctic America. Wintering in S. America.

One has been shot in Cornwall (1871), and another in Leeds Museum is said to have been killed previously (1854-5) in Lincolnshire.

GENUS CLXXX. GLOTTIS, *Koch (1816).*

Differs chiefly from *Totanus* in having the bill very slightly *up-curved.*

316. Glottis canescens (Gmel.). GREENSHANK.

Hab. Northern Palæarctic region.

Adult: feathers of back and wings dark brown, with blackish centres and greyish-white edges; head and nape greyish-white, with blackish streaks; rump white; tail white, irregularly barred with dusky-brown; under parts white, spotted with blackish on throat and breast and slightly barred on sides; bill blackish, paler at base; tarsi greyish-green. In winter paler and greyer above and almost unspotted white below. Length 11·75; wing 7·25.

A regular but not abundant migrational visitor; sometimes remaining through the winter. Some numbers remain to breed in the Highlands and Hebrides.

GENUS CLXXXI. MACRORHAMPHUS, *Leach (1816).*

317. Macrorhamphus griseus (Gmel.). RED-BREASTED SNIPE.

Hab. Arctic America. In winter reaching S. America.

About a dozen have been killed in England, two in Scotland and one in Ireland.

317a. M. griseus scolopaceus (Say). LONG-BILLED SNIPE.

Hab. Western Arctic America.

An example received by Mr. Coburn, of Birmingham,

from Tipperary, 11th October, 1893, was referred by Professor Newton to this variety.

GENUS CLXXXII. LIMOSA, *Brisson* (1760).

Bill very long, slightly *up-curved*, compressed and tapering, but a little dilated at point; upper mandible longer than lower. Bill is longer in the female than in the male. Tibia chiefly naked; tarsus long, slender; outer toe united to middle at base; hind toe moderate.

318. Limosa lapponica (L.). BAR-TAILED GODWIT.

Hab. Arctic Europe and N.W. Siberia.

Adult: feathers above rufous-brown, with blackish centres; wing-coverts snd secondaries edged with dull white; rump white with dusky streaks; tail dusky-brown, barred with white; chin, throat, neck and under parts chestnut-red, streaked with black on sides of breast. In winter feathers greyish-brown above, with blackish centres and greyish-white margins; rump white; tail chiefly dusky-grey; upper tail-coverts white, with dusky bars; below white, with some dusky streaks on sides of throat and breast. Length: male 12·50, wing 8·00; female 13·50, wing 8·20; bill flesh-colour, dusky towards tip; tarsi bluish-slate.

A regular migrational visitor to our coasts.

319. Limosa limosa (L.). BLACK-TAILED GODWIT.

Hab. Northern Europe and Western Asia.

Adult: distinguished from *L. lapponica* by tail being *black*, with the base white only; greater wing-coverts and chief part of secondaries are white, forming a conspicuous bar; red tint below is restricted to cheeks, throat, and breast, and is more barred with black; belly whitish with dusky bars. In winter greyish-brown above, darker on rump; upper throat and belly whitish; lower throat, breast and sides brownish-grey; tail and wings as in spring. Length: male 13·00, wing 8·75; female 13·75, wing 9·00.

Now merely a regular migrational visitor, but much less common than *L. lapponica*.

Genus CLXXXIII. NUMENIUS, *Brisson* (*1760*).

Bill somewhat as in *Limosa*, but *much decurved*. Feet as in *Limosa*.

320. Numenius arquatus (Linn.). Curlew.

Hab. North Europe. In winter reaching Africa.

Adult: feathers above brownish-buff, with narrow dusky centres; primaries blackish; rump and upper tail-coverts pure white, somewhat streaked on former and barred on latter with dusky-brown; tail buffish-white barred with dusky-brown; lower throat, breast and sides like upper parts; upper throat, belly, and under tail-coverts white, with a few streaks on latter; bill very pale brown, dusky at tip; tarsi dusky-grey. In winter paler above and below; rump almost unstreaked; under parts white, with fewer markings. Length: male 17·50, wing 11·00; female 19·50, wing 12·00.

Breeds on moors and waste lands throughout the British Isles excepting the south-east of England. It is particularly common on our coasts in winter. Eggs: 4; varying from olive-green to olive-brown, blotched and spotted with dark brown; 2·80 by 1·95.

321. Numenius phæopus (Linn.). Whimbrel.

Hab. Northern Europe.

Adult: a miniature Curlew in appearance, but differs in having top of head dark brown on each side with a buffish-white stripe in the middle, there being also another whitish stripe over each eye. Length: male about 13·00, wing 9·50; female slightly larger.

A common visitor in spring and autumn to our coasts, a proportion remaining throughout the summer. It is only known to breed in the Shetlands, Orkneys, and some of the Outer Hebrides.

322. Numenius borealis (Forster). ESQUIMAUX CURLEW.

Hab. Arctic America. Wintering in S. America.

A rare straggler, not more than eight examples having been obtained in the British Isles.

ORDER GAVIÆ.

Family Laridæ.

Sub-Family Sterninæ.

GENUS CLXXXIV. **STERNA**, *Brisson* (*1760*).

Bill longer than head, compressed, tapering to a sharp point, nearly straight, but upper mandible gently decurved. Nostrils placed at about middle of bill; wings long and pointed; tail more or less deeply forked; tarsus short; feet webbed; hind toe small and elevated.

323. Sterna fluviatilis, Naum. COMMON TERN.

Hab. Palæarctic region; also Eastern North America.

Adult: head and nape black; mantle pale grey; rump and tail white, with grey outer margins to the long tail-eathers; below white, with a pinkish-grey tinge on the breast; bill and feet orange-red. In winter crown is nearly white. Length 13·00. Young: crown and nape mottled with brownish-black, and mantle with greyish-brown; bill and feet yellowish, becoming dusky during first winter.

Common from May to September on our coasts, excepting the north of Scotland. Eggs: 2 or 3; varying from pale stone-colour to brownish-buff, blotched and spotted with dark brown and grey; 1·65 by 1·15.

324. Sterna macrura, Naum. ARCTIC TERN.

Hab. Circumpolar region. In winter reaching Equator.

Differs from *S. fluviatilis* in having the upper parts darker, and under parts pale ash-grey; bill and feet dark red; tarsi and wing somewhat shorter.

Common in spring in the Shetlands, Orkneys, Hebrides, all around Scotland, and on the N.E. coast of England; also the north, west and south of Ireland. Eggs: often with a more olive or greenish tint than the Common Tern's, and frequently more boldly blotched or zoned; 1·60 by 1·10.

325. Sterna dougalli, Montagu. ROSEATE TERN.

Hab. Temperate and tropical regions of globe.

Rather smaller than *S. fluviatilis*, but with a proportionately longer and entirely grey tail and of a paler grey on mantle, while under parts are pure white tinged with pink; bill black; feet red. Length 14·00; wing 9·00.

Formerly breeding rather commonly on our coasts, but it appears now to only breed on the Farne Islands.

326. Sterna minuta (Linn.). LITTLE TERN.

Hab. Palæarctic region, eastward to North India.

Adult: forehead white; rest of head and nape black; mantle and wings pale grey, primaries darker and margined with white on inner webs; rump, tail, and under parts white; bill orange with a blackish tip; feet orange. Length 8·00; wing 6·70. Young: head, nape, and mantle mottled with blackish-brown; bill and feet brown.

Moderately common from May to September on most parts of our coasts. Eggs: 2 or 3; stone-colour with dark brown and grey spots, but not blotched; 1·30 by ·95.

327. Sterna caspia, Pall CASPIAN TERN.

Hab. Almost cosmopolitan.

A rare straggler to east coast of England, where 16 or 17 have been obtained and others seen. Has also occurred in Dorset and Hants.

328. Sterna anglica, Montagu. GULL-BILLED TERN.

Hab. Southern Palæarctic and Nearctic regions.

A rare straggler to England, scarcely more than twenty examples having been obtained. The reported Irish specimen turns out to be an immature Arctic Tern.

329. Sterna cantiaca, Gmel. SANDWICH TERN.

Hab. Temperate Europe, N. Africa and S.W. Asia.

Adult: head and nape black; mantle and wings pale grey, primaries darker and margined with white on inner webs; rump, tail and under parts white, latter with a pink tinge. In winter black is chiefly confined to nape. Length 14·75; wing 12·00.

A summer visitor in small numbers to the English coasts and to several localities on both sides of Scotland; in Ireland only known to breed in one locality near Ballina, Co. Mayo. Eggs: 2 or 3; ranging from cream-colour to warm buff, spotted variably with reddish-brown of two shades and grey; 2·00 by 1·45.

330. Sterna fuliginosa, Gmel. SOOTY TERN.

Hab. Tropical regions of the globe.

One was shot in 1852 near Burton-on-Trent, a second in 1869 in Berkshire and a third in 1885 near Bath.

331. Sterna anæstheta, Scop. SCOPOLI'S SOOTY TERN.

Hab. Tropical seas of the globe.

One was taken at the mouth of the Thames in 1875 (Zool. 1877, p. 213; P.Z.S., 1877, p. 43).

GENUS CLXXXV. HYDROCHELIDON, *Boie (1822).*

Bill scarcely longer than head. Tail rather short, and but little forked. Feet with the webs more deeply scalloped than in *Sterna*.

332. Hydrochelidon hybrida (Pall.). WHISKERED TERN.

Hab. Southern Palæarctic and Oriental regions; also Africa and Australia.

Has occurred in Dorset (1836), Dublin Bay (1839), Yorks (1842), Norfolk (1847), Cornwall (1851), and near Plymouth (1865).

333. Hydrochelidon leucoptera (Schinz). WHITE-WINGED BLACK TERN.

Hab. Southern Palæarctic region.

An irregular straggler to England in spring; has also occurred five times in Ireland.

334. Hydrochelidon nigra (Linn.). BLACK TERN.

Hab. Europe, north to Baltic; also N. Africa.

Adult: whole plumage sooty slate-grey, turning to blackish on head and nape; vent white; bill blackish; feet dull brownish-red. In winter, forehead, nape and chin are dull white, and under parts are much mottled with white. Length 9·00, wing 8·50. Young: like adults in winter, but almost entirely white below and mottled with brown on mantle.

Formerly breeding commonly in East Anglia, but now chiefly known as a migrational visitor in small numbers to our coasts.

GENUS CLXXXVI. ANOUS, *Stephens (1825).*

335. Anous stolidus (Linn.). NODDY.

Hab. Tropical regions of the globe.

Two were shot off Co. Wexford about 1830.

Sub-Family Larinæ.

GENUS CLXXXVII. PAGOPHILA, *Kaup (1829).*

336. Pagophila eburnea (Phipps). IVORY GULL.

Hab. Circumpolar regions.

Scarcely more than thirty examples have been obtained in the British Isles.

GENUS CLXXXVIII. RISSA, *Stephens (1825).*

Bill shorter than head, rather stout, compressed, upper mandible arched and decurved towards tip; tail moderate, nearly even; tarsus short; hind toe obsolete; front toes fully webbed.

337. Rissa tridactyla (Linn.). KITTIWAKE.

Hab. Northern Palæarctic and N.E. Nearctic regions.

Adult: mantle and wings grey; outer primaries black on terminal portions; rest of plumage pure white; bill yellowish; feet blackish. In winter nape is slate-grey.

Length 14·25, wing 12·00. Young; mantle and wing-coverts mottled with blackish-brown; nape dusky; tail with a terminal band of dusky-brown; bill black; feet brownish.

Abundant on our coasts all the year, breeding commonly on the rocky portions. Eggs: 2 or 3; varying from stone-colour to olive-buff, spotted and blotched with chestnut brown and lilac-grey; 2·10 by 1·55; laid in a nest of seaweed on rock-ledges.

Genus CLXXXIX. LARUS, *Linnæus (1766).*

Bill not so long as head, strong, upper mandible decurved at tip, gonys with a prominent angle. Tail moderate, square. Tarsus moderately long; hind toe present, but small and elevated; front toes fully webbed.

338. Larus glaucus, Fabr. Glaucous Gull.

Hab. Northern Palæarctic and N.E. Nearctic regions.

An irregular winter visitor, mostly occurring on the Scotch coasts and the east side of England.

339. Larus leucopterus, Faber. Iceland Gull.

Hab. Arctic America and Greenland. In winter westward to Iceland and N.W. Europe.

Another white-winged species, but distinguished from the last by its smaller size. Length 20·00 to 21·00; wing 16·00. Occurs irregularly in winter.

340. Larus argentatus, Gmel. Herring-Gull.

Hab. Western Europe and N.E. America.

Adult: mantle and wings light grey; secondaries with white tips; outer primaries black with white tips and sub-terminal spots and grey margins to basal portions of inner webs; rest of plumage white; bill yellow, tip reddish; feet yellowish flesh-colour. In winter crown and nape exhibit dusky streaks. Length of male 22·00, wing 17·00; female slightly smaller. Young: chiefly brown on mantle and mottled with brown on rest of plumage.

Common and breeding abundantly in very many localities. Eggs: 3; pale olive-brown, with blotches and spots of dark brown; 2·90 by 2·00; laid in a nest of grasses on rock-ledges, or occasionally on grassy islets.

341. Larus fuscus, L. LESSER BLACK-BACKED GULL.

Hab. Europe and N. Africa.

Adult: mantle and wings dusky slate, often almost black; first one or two primaries with sub-terminal white patches and most of primaries and secondaries slightly tipped with white; rest of plumage white; bill yellow, tip red; feet yellow. In winter head and nape show dusky streaks. Length of male 21·00, wing 16·25. Young: much darker than young of *L. argentatus.*

Breeds abundantly on the most rocky coasts of our islands; it is more generally diffused in winter. Eggs: 3; similar to those of *L. argentatus* but more variable; 2·80 by 1·95; laid in May in a nest of grass and seaweed placed on grassy islets, "stacks," etc.

342. Larus marinus, L. GREAT BLACK-BACKED GULL.

Hab. Northern Europe and N.E. America.

Distinguished from *L. fuscus* by its larger size. Length of male 27·00, wing 20·00. Breeds commonly everywhere around Scotland and less so down the west coast of England and Wales to the Scilly Isles, while colonies are found all round the Irish coasts. Birds which are not breeding frequent the English coasts at all seasons. Eggs: 2 or 3; like those of *L. argentatus* but larger; 3·00 by 2·15.

343. Larus canus, Linn. COMMON GULL.

Hab. Northern Palæarctic region.

Adult: mantle and wings grey, excepting outer primaries which are black with white tips and sub-terminal patches; rest of plumage white; bill greenish-yellow, orange at tip; feet greenish-yellow. In winter head and neck are streaked with dusky-brown. Length 17·00; wing 14·00. Young: upper *and under* wing-coverts mottled with dark brown; tail with a terminal dusky-brown band; primaries dusky-brown.

Breeds commonly around Scotland and its islands and down the west side of Ireland, but is only known in England as a common winter visitor. Eggs: 3; usually light olive brown spotted and blotched with umber and with underlying grey marks; 2·25 by 1·60; laid in a nest of seaweed and grass on grassy islets.

344. Larus ichthyaetus, Pall. GREAT BLACK-HEADED GULL.

Hab. N.E. Africa, the Levant and S.W. Asia.

An example in the Exeter Museum was shot off Exmouth in 1859.

345. Larus melanocephalus, Natterer. MEDITERRANEAN GULL.

Hab. Shores of the Mediterranean.

One in the Natural History Museum is said to have been shot near Barking Creek, Essex (1866), a second has since been obtained on Breydon Broad (1886).

346. Larus ridibundus, L. BLACK-HEADED GULL.

Hab. Temperate Europe and Asia.

Adult: head, cheeks, chin and upper throat dark brown; mantle and wing-coverts pale grey; primaries white in centre with blackish margins to both webs; rest of plumage white. In winter "black" head is lost. Length 14·50; wing 12·00. Young: wing-coverts mottled with dark brown; tail with a terminal band of dusky-brown.

Quite the commonest of its genus, breeding all around our coasts. Eggs: 3; usually light olive-brown, occasionally olive-green, variably blotched and spotted with umber; 2·20 by 1·50; laid in a nest of sedges, grass, etc. on the ground in marshes.

347. Larus minutus, Pall. LITTLE GULL.

Hab. Eastern Europe and temperate Asia. In winter migrating southward and irregularly westward.

An irregular but not uncommon winter visitor to the east side of Great Britain, and also the south coast; much less frequent on the west, and a very rare visitor to Ireland.

348. Larus philadelphia (Ord). BONAPARTE'S GULL.

Hab. Arctic America (chiefly the interior).

Four examples have been taken in England, one in Scotland, and one in Ireland.

GENUS CXC. RHODOSTETHIA, *Macgillivray* (*1842*).

349. Rhodostethia rosea, Macg. ROSS'S GULL.

Hab. Polar regions (breeding range unknown).

An example in Leeds Musuem is said to have been shot at Tadcaster, Yorks, in 1846 or 1847.

GENUS CXCI. XEMA, *J. Ross* (*1819*).

350. Xema sabinii (J. Sabine). SABINE'S GULL.

Hab. Arctic America and N.E. Asia.

A scarce and irregular autumn visitor to our coasts. In some winters several specimens are obtained, but it is by no means of annual occurrence.

Sub-Family Stercorariinæ.

Genus CXCII. STERCORARIUS, *Brisson (1760)*.

Bill strong, tip of upper mandible dilated, decurved and somewhat hooked. Nostrils placed toward extremity of bill, narrow and oblique. Toes with large hooked claws.

351. Stercorarius catarrhactes (Linn.). Common Skua.

Hab. N.W. Europe and Eastern Arctic America.

Adult: feathers above and below chiefly dark brown with rufous margins; primaries *white* on basal portion; bill and feet blackish. Length about 22·00; wing 16·00.

Small protected colonies exist in the Shetlands, but it does not breed elsewhere, neither is it very plentiful on our coasts in winter; it is rarely recorded from Ireland.

352. Stercorarius pomatorhinus (Temm.). Pomatorhine Skua.

Hab. Arctic regions of Asia and America.

An irregular visitor in autumn and winter, occurring annually in varying numbers on the east side, but more rarely on the west and in Ireland. It is larger than the next species, and the adult has the two elongated middle tail-feathers broad and twisted into a *vertical* position.

353. Stercorarius crepidatus (Gmel.). Richardson's Skua.

Hab. Northern Palæarctic and Nearctic regions.

There is a light and a dark phase of this Skua, extremes of former having throat and breast white, and extremes of latter having whole plumage dusky brown, but the two constantly interbreed, and every intermediate stage is to be found. In adults two middle tail-feathers are elongated and tapering, but much shorter than in the next species. Length 19·00; wing 13·50; shafts of *all* the primaries are white at all ages.

Breeds on the Shetlands, Orkneys, Hebrides and a few spots in the Highlands; also of frequent occurrence in winter on all our coasts. Eggs: 2; olive-green, blotched and spotted with deep brown; 2·30 by 1·60.

334. Stercorarius parasiticus (Linn.). LONG-TAILED SKUA.

Hab. Circumpolar regions.

Adult: smaller and more slender than *S. crepidatus*, and with shafts of 2 outer primaries only white; upper parts and belly greyish-brown; head nearly black; neck dull yellow; breast white. Length to tip of tail 22·00; wing 11·75. Female has a rather shorter tail.

An irregular autumn visitor, sometimes occurring in fair numbers on the east side of Great Britain.

ORDER TUBINARES.

Family Procellariidæ.

GENUS CXCIII. PROCELLARIA, *Linnæus* (*1766*).

Bill moderate, straight, excepting the horny nail at tip, which is much decurved. Nostrils tubular, placed upon upper surface of bill. Wings long, narrow. Tail moderate, nearly even. Tarsi moderate; hind toe obsolete.

355. Procellaria pelagica, Linn. STORM-PETREL.

Hab. Atlantic coasts of Europe; also Mediterranean.

Adult: plumage black, excepting upper tail-coverts and sides of rump, which are conspicuously white; bill and feet black. Length 5·25; wing 4·65.

Common on all our coasts, particularly during migration. Breeds numerously on the coasts of Scotland, Ireland, Wales and the Scilly Isles. A single egg is laid in a hole burrowed in the ground or under boulders; it is pure white, sometimes speckled with rufous; 1·10 by ·85.

GENUS CXCIV. **OCEANODROMA,** *Reichenbach (1853).*

356. Oceanodroma leucorrhoa (Vieill.). LEACH'S PETREL.

Hab. Islands of North Atlantic and North Pacific.

Chiefly known as an irregular visitor to our coasts, especially the eastern side. It has, however, long been known to breed at St. Kilda, and has more recently been found breeding in the outer Hebrides and on the Blasquets off the coast of Kerry.

Distinguished from *P. pelagica* by its deeply forked tail and larger size; length 6·75; wing 6·00.

357. Oceanodroma cryptoleucura.

Hab. Islands of the South Atlantic.

One was picked up on the beach near Dungeness, December 5th, 1895.

GENUS CXCV. **OCEANITES,** *Keyserling & Blasius (1840).*

358. Oceanites oceanicus, Kuhl. WILSON'S PETREL.

Hab. Islands of South Atlantic and South Pacific.

A rare straggler to England, nine or ten examples having been obtained and others seen. One is said to have been taken in Ireland in 1840, while two more were obtained in October, 1891.

GENUS CXCVI. **PUFFINUS,** *Brisson (1760).*

Bill slightly longer than head, somewhat slender, both mandibles decurved at tip, particularly upper one.

359. Puffinus anglorum (Tm.). MANX SHEARWATER.

Hab. Coasts and islands of North Atlantic.

Adult: above sooty-black; below white, with some greyish-brown mottlings on sides of upper breast; bill dusky-brown; feet pale dull yellow. Length 13·25; wing 9·50.

Breeds rather commonly along west side of Great Britain, also in the Shetlands, Orkneys and Hebrides. There are many colonies in the west and south-east of Ireland. A single white egg is laid in a burrow in the ground; 2·40 by 1·70.

360. Puffinus griseus (Gm.). SOOTY SHEARWATER.

Hab. Islands of southern hemisphere. Frequenting coasts of northern hemisphere during summer.

Constantly confounded with the next, but it has been proved to occur casually on many parts of our coasts.

361. Puffinus major, Faber. GREAT SHEARWATER.

Hab. Probably as *P. griseus.* Common on coasts of North Atlantic during northern summer.

A fairly regular visitor, but in very varying numbers. Adult is greyish-brown above, darker on head, wings and tail, and mottled with white on tail-coverts; below whitish; bill dark brown; feet "pinkish-white in life." Length 16·00; wing 13·00.

362. Puffinus obscurus (Gm.). DUSKY SHEARWATER

Hab. West Indies, Bermudas, Canaries and Madeira.

One was obtained off the coast of Kerry in 1853, and a second in Suffolk in 1858.

GENUS CXCVII. FULMARUS, *Stephens* (1826).

Bill slightly shorter than head, stout, strong; nail large and much decurved; tip of under mandible bent, gonys with a prominent angle; nasal tube very prominent.

363. Fulmarus glacialis (Linn.). FULMAR.

Hab. Arctic and sub-Arctic regions of North Atlantic.

Adult: mantle and tail light grey; primaries dusky; rest of plumage usually white (specimens of dark phase being rare on our coasts); bill pale yellow, dusky on nasal tube; feet slate. Length 17·00; wing 12·50.

Breeds in St. Kilda, Foula, and Papa Stour in the Shetlands, and probably some of the outer Hebrides. To the coasts of England it is a scarce winter visitor, while only fifteen or sixteen examples have been recorded from Ireland. A single white egg is laid on ledges of cliffs; 2·90 by 2·00.

GENUS CXCVIII. ŒSTRELATA, *onaparte* (*1856*).

364. Œstrelata hæsitata (Kuhl). CAPPED PETREL.

Hab. Probably the South Atlantic.

An example was captured in 1850 near Swaffham, Norfolk (Zool. 1852, p. 3691).

GENUS CXCIX. BULWERIA, *Bonaparte* (*1842*).

365. Bulweria columbina (Moquin-Tandon). BULWER'S PETREL.

Hab. Canaries and Madeira.

An example was found dead near Tanfield, Yorkshire, on May 8th, 1837 (P. Z. S., 1887, p. 562).

ORDER PYGOPODES.

Family Colymbidæ.

GENUS CC. COLYMBUS, *Linnæus* (*1766*).

Bill as long as head, straight, strong, somewhat compressed, sharply pointed; nostrils lateral, near the base

Wings rather short; tail very short. Legs short, placed very much behind; tarsus compressed; front toes fully webbed; hind toe small.

366. Colymbus glacialis, Linn. GREAT NORTHERN DIVER.

Hab. N.W. Europe and N.E. America.

Adult: above black, spotted in transverse rows with white; throat and neck black, upper throat with a violet gloss and a band of ten or twelve short longitudinal streaks; lower throat glossed with green and with a band of about eighteen streaks; below white, streaked with black on upper breast; bill black; iris red. Length about 29·00; wing 13·50. Young: feathers above dusky with greyish margins; below whitish; bill pale brown.

Of fairly common occurrence on our coasts from autumn until late spring. It is believed to breed in some of the Shetlands.

367. Colymbus adamsi, Gr. YELLOW-BILLED DIVER.

Hab. Circumpolar regions, breeding far north.

One was shot near Lowestoft in 1852, another (believed to be obtained in the county) is figured in Babington's "Birds of Suffolk," a third, killed in Northumberland, is in the Newcastle Museum, while it has recently been proved (Zool., 1896, p. 14) that a fourth was killed on Hickling Broad in 1872 by the late E. T. Booth.

368. Colymbus arcticus, L. BLACK-THROATED DIVER.

Hab. Northern Palæarctic and Nearctic regions.

Adult: above blackish, barred on mantle and back and spotted on wings with white; head and nape grey; chin and throat black, with a rich purplish sheen, and with a band of white streaks below chin; below white, streaked with black on sides of upper breast; bill black;

iris red. Length about 24·00; wing 11·50. Young: like young of *C. glacialis*, but smaller.

Breeds rather commonly on the lochs in many parts of the Highlands, also in the Orkneys and Hebrides, but to the rest of our coasts it is a rare winter visitor. Eggs: 2; olive-brown, sparingly marked with blackish-brown; 3·00 by 2·00; on ground close to water's edge.

369. Colymbus septentrionalis, Linn. RED-THROATED DIVER.

Hab. Northern Palæarctic and Nearctic regions.

Adult: above marked with small longitudinal spots of white, not bars; cheeks and sides of neck grey; upper throat red; below white. In winter red throat is lost. Length 21·00; wing 10·75. Young: at first spotted above with white; later with the feathers dusky, but edged distinctly with white.

Breeds throughout the Highlands, also in the Shetlands, Orkneys, and Hebrides, and of recent years in Co. Donegal. It is of common occurrence on our coasts in winter. Eggs: 2; smaller than those of *C. arcticus*; 2·75 by 1·85.

Family Podicipedidæ.

GENUS CCI. PODICIPES, *Latham (1787).*

Bill much as in *Colymbus*; toes flattened and furnished on inner sides with broad membranes.

370. Podicipes cristatus (Linn.). GREAT CRESTED GREBE.

Hab. Temperate portions of Old World.

Adult: crown, crest and upper parts dark brown; secondaries white; chin and cheeks white, the feathers of both elongated and hair-like, with the tips chestnut edged with black; neck and upper parts silky and greyish-white. Length 20·00; wing 8·25. In winter crest and elongated feathers below head are lost.

Breeds locally on broads, meres, etc., throughout the eastern, midland and western counties of England, and even in South Scotland; also in many parts of Ireland. Eggs: 4 or 5; at first white; shell rough; 2·15 by 1·55; laid on a mass of wet sedges, etc., on surface of water.

371. Podicipes griseigena (Bodd.). Red-necked Grebe.

Hab. Eastern and Northern Europe.

An irregular winter visitor, occurring chiefly on east side, only five or six examples having been recorded from Ireland. It is smaller than *P. cristatus*, and feathers of cheeks and upper throat are not elongated, while head and crest are blacker and front of neck is rufous in adults.

372. Podicipes auritus (Linn.). Sclavonian Grebe.

Hab. Northern Palæarctic and Nearctic regions.

Another winter visitor; of more regular occurrence than the last. In winter adults are like immature birds which lack the elongated ear-tufts, crest and cheek-feathers of spring, and bear more resemblance to last species; *three outer secondaries* are, however, dusky-brown. Length 12·00; wing 5·75.

373. Podicipes nigricollis, Brehm. Eared Grebe.

Hab. Southern Palæarctic region and all of Africa.

A straggler to England in spring and casually in autumn; it may even have bred in Norfolk. It is not often recorded from Scotland or Ireland.

GENUS CCII. TACHYBAPTES, *Reichenbach* (*1851*).

374. Tachybaptes fluviatilis (Tunst.). LITTLE GREBE.

Hab. Temperate and tropical regions of Old World.

Adult: above deep brown; secondaries showing scarcely any white; chin blackish; cheeks, throat, and upper breast rufous; belly greyish-white; sides dark brown; bill horn-brown, greenish at base on each side; iris brownish-red; feet greenish. Length 8·50; wing 4·25. In winter paler and with the chin whitish. Young: chiefly brownish-white below.

Common everywhere on ponds, lakes and streams. Eggs: 4 to 6; much smaller than *P. cristatus*; 1·50 by 1·00.

Family Alcidæ.

GENUS CCIII. ALCA, *Linnæus* (*1766*).

Bill rather large, upper mandible with the terminal part arched, enlarged and much decurved, lower correspondingly bent, base of both mandibles clothed with short feathers. Nostrils marginal, narrow, situated in middle of bill. Hind toe absent; front toes fully webbed.

375. Alca torda, Linn. RAZORBILL.

Hab. Coasts and islands of North Atlantic.

Adult: above black, glossed with green; throat dark brown; below white; greater wing-coverts slightly tipped with white; bill black, with a white groove on each mandible and a white line in front of eye. In winter throat is white and upper parts are slightly duller. Length 15·50; wing 7·50. Young: resembles adults in winter but white grooves on bill are absent and white in front of eye is scarcely visible.

Breeds in large numbers on the rocky portions of our coasts, depositing in May a single egg on a rock-ledge or in a crevice ; it is less pyriform than that of the Common Guillemot and greyish (or brownish) white, boldly blotched and spotted with brownish-black ; 2·80 by 1·95.

376. Alca impennis, Linn. GREAT AUK.

Hab. Formerly coasts and islands of North Atlantic. Now apparently quite extinct.

GENUS CCIV **URIA,** *Brisson (1760).*

Bill about as long as head, straight, pointed, strong and compressed ; tip of upper mandible slightly decurved and indented ; nostrils basal, lateral.

377. Uria troile (Linn.). COMMON GUILLEMOT.

Hab. Coasts and islands of North Atlantic.

Adult : upper parts, chin and throat deep uniform brown; below white ; bill and feet blackish, webs greenish. In winter throat is chiefly white. Length 16·00 ; wing 7·75. The Ringed Guillemot (*U. lacrymans* of Macgillivray and Yarrell) is now admitted to be merely a variety of *U. troile.*

Breeds very abundantly on all our rocky coasts. Egg : very pyriform in shape ; varying from greenish-white to sea-green or even reddish-brown, streaked and heavily blotched with blackish-brown ; 3·10 by 2·00.

378. Uria bruennichi, Sab BRÜNNICH'S GUILLEMOT.

Hab. Arctic regions ; almost circumpolar.

A very rare straggler, only two or three examples having apparently been actually obtained on our coasts. The late A. G. More excluded it from the Irish list.

379. Uria grylle (Linn.) BLACK GUILLEMOT.

Hab. Coasts and islands of North Atlantic.

Adult : black, with a large white patch on the wing-coverts ; bill black ; iris hazel ; feet red. In winter mottled and barred above with white; rump and under parts chiefly white. Length 12·50; wing 6·50. Young: similar to adults in winter plumage, but bill is dusky-slate; iris darker; feet dark brown.

Breeds on the rocky coasts of Scotland and Ireland, but on the English coasts very few birds are to be found even in winter. Eggs: 2; greyish-white, spotted and blotched with deep brown and ash-grey; 2·35 by 1·60

GENUS CCV. MERGULUS, *Vieillot (1816)*.

380. Mergulus alle (Linn.). LITTLE AUK.

Hab. Coasts of North Atlantic, within Arctic Circle.

A very irregular winter visitor, but sometimes occurring in large numbers, particularly along the east coast. It is easily recognised by its small size, black and white plumage, and short, stout bill. Length 7·75; wing 4·50.

GENUS CCVI FRATERCULA, *Brisson (1760)*.

Bill higher than long (in breeding season), much compressed and transversely grooved; nostrils basal, marginal.

381. Fratercula arctica (Linn.). PUFFIN.

Hab. Coasts and islands of North Atlantic.

Adult: crown, nape, a collar round throat, and upper parts black, lighter on wings; sides of head, cheeks, chin and under parts white; bill variegated with red, yellow and grey. In winter the outer sheath of bill is lost, its size being proportionately reduced. Length about 11·00; wing 6·00. Young: duller and with face grey; bill as in adults in winter.

Breeds in great numbers on all our coasts with the exception of the portion from the Humber to Hants. A single egg is laid in crevices of rocks or in a burrow in the soil; at first white; surface rough; 2·20 by 1·60.

APPENDIX.

PROVISIONALLY EXCLUDED SPECIES,

All of which may *possibly* be genuine visitors, but whose recorded occurrences are either insufficiently authenticated or somewhat improbable. Position in foregoing list is indicated.

6-7. **Turdus sibiricus** (Siberia).
81-2. **Tachycineta bicolor** (N. America).
83-4. **Progne purpurea** (N. America).
106-7. **Emberiza cioides** (Siberia).
112-13. **Agelæus phœniceus** (N. America).
112-13. **Sturnella magna** (N. America).
112-13. **Scolecophagus ferrugineus** (N. America).
114-15. **Pyrrhocorax pyrrhocorax** (Central Europe).
127-8. **Melanocorypha calandra** (S. Europe).
135-6. **Picus martius** (Northern Europe).
135-6. **Dendrocopus villosus** (N. America).
135-6. **Dendrocopus pubescens** (N. America).
137-8. **Picoides tridactylus** (N. Europe).
138-9. **Colaptes auratus** (N. America).
164-5. **Buteo desertorum** (Africa and Asia).

174-5. **Elanus cæruleus** (Africa).
209-10. **Branta canadensis** (N. America).
212-13. **Cygnus americanus** (N. America).
212-13. **Cygnus buccinator** (N. America).
231-2. **Clangula islandica** (Iceland).
252-3. **Caccabis petrosa** (N.W. Africa).
261-2. **Porzana carolina** (N. America).
264-5. **Porphyrio porphyrio** (S.W. Europe).
264-5. **Porphyrio smaragdonotus** (Africa).
267-8. **Grus virgo** (S. Europe and Africa).
267-8. **Balearica pavonina** (N. Africa).
288-9. **Phalaropus wilsoni** (N. America).
309-10. **Tringoides macularius** (N. America).
315-6. **Totanus stagnatilis** (S. Europe).
345-6. **Larus atricilla** (N. America).

ERRATA.

Page 25, line 24, *for* "Harve" *read* "Harvie."
" 41 " 11 " "51" *read* "51a."

PHILIP JOHNS & CO., LTD., 6, WORSHIP STREET, E.C.

JOHN WHELDON & CO.,

58, GREAT QUEEN ST , LONDON, W.C

The following List is a selection from our stock of Second-hand Ornithological Works. For others, including papers, &c., please apply.

AMERICA (GOULD, John). Odontophorinæ, or Partridges of America, 32 *coloured plates, with full descriptions*, 1 vol, imp. folio. *half morocco extra, gilt edges*, £10 10s

ALDRIDGE (W.) A Gossip on the Birds of Norwood and Crystal Palace District, *with* 13 *full-page woodcuts*, 8vo, *cloth*, 1s, *a few copies only for sale.*

BAKER (T. B. L.) An Ornithological Index, arranged according to the Synopsis Avium of Mr Vigors, 8vo., 1s 1835.

BEWICK (Thomas). History of British Birds, 2 vols, 1805; ditto Quadrupeds, 1807, in 3 vols, extra large paper, newly bound in whole brown *calf gilt, yellow edges*, roy. 8vo, £5 15s 1805

An exceptionally clean and fine set of this delightful work

BOOTH (E. T.) Rough Notes on the Birds observed during Twenty-five Years' Shooting and Collecting in the British Isles, in 3 vols, *containing* 116 *beautifully coloured plates*, folio, *a subscriber's copy, in boards*, £18 18s

BUTLER (A. G.) British Birds' Eggs, a Hand-Book of British Oology, 38 *beautiful coloured plates*, 8vo, *cloth*, £1 5s 1886

BUTLER (E. A.) A catalogue of the Birds of Sind, Cutch, Ka'thia'wa'r, North Gujarat and Mount Aboo, including every species known to occur in that tract of country up to date, with references showing where each species is described and locality marking its distribution 80 years, etc., 2s 6d 1879

CORY (Chas. B.) The Birds of the Bahama Islands, containing many Birds new to the Islands, and a number of undescribed Winter Plumages of North American Birds, *illustrated with* 8 *large full page plates*, 4to, Roxburghe binding, 10s 6d (pub 25s) Boston

DIXON (Rev. E. S.) The Dovecote and Aviary, being Sketches of the Natural History of Pigeons *woodcuts*, 12mo, *cloth*, 2s 6d (pub 7s 6d) 1851

ELLIOT (D. G.) North American Shore Birds, *illustrated with* 74 *fine full page plates drawn for this work by* E. SHEPPARD, 8vo., *cloth* 8s 6d 1895

ELLIOT'S Tetraoninæ, or Grouse Family, 5 parts in 4, forming 1 vol complete, 25 *coloured plates of Birds, and 2 of Eggs, drawn from nature by the Author, the figures life size,* imp. folio, *boards*, £7 10s New York, 1864—65

EYTON (T. C.) History of the Rarer British Birds, *illustrated largely by most highly finished woodcuts of birds and exquisite tail pieces*, 8vo, *newly bound in half green mor. extra, top edge gilt*, 10s 6d 1836

Ditto, another copy cloth, quite clean and uncut, 6s 6d

FALCONRY Il. Falconiere di Jacopo Augusto Tuano primo presidente del parlemento di parigi, e consigliere intimo di Arrigo Quarto dall esametro latino all' endecasillabo Italiano Tras Ferito ed interpretato coll' uccellatura a vischio di Pietro Angelio Bargeo, *with numerous engravings including a fine frontispiece*, 4to, *calf, rebacked, scarce*, £1 15s 1735

GERINI (Giovanni) Storia Naturale degli uccelei trattata con metodo e adornata di figure intagliate u rame e miniate al naturale, *illustrated with 600 hand coloured plates, of birds drawn in their most natural positions*, in 5 vols, folio, *half calf*, £24 1767—76

The above copy of this grand old work is in excellent preservation.

GRAY (G. R.) Fasciculus of the Birds of China, 12 *beautifully coloured plates*, imp. 4to., *boards*, scarce, 15s 1871

GRAY'S. List of the Genera of Birds, with an Indication of the Typical Species of each Genus, 8vo., First Edition, *cloth*, 3s 6d scarce 1840

——— Ditto 2nd Edition *cloth* 3s 6d

GRAY (R.) and (Anderson T.) Birds of Ayrshire and Wigtonshire, describing the Bird-Life of the Coasts and Locks, the Moors, Glens and Valleys, where inhabit many Birds of great variety, *plate* 8vo 3s 6d scarce 1869

HARVIE BROWN and others. 5th Report on the Migration of Birds, scarce 3s 6d post free 3s 9d Edinburgh, 1883

HARTING (J. E.) A Hand book of British Birds, showing Distribution of the Resident and Migratory Birds in the British Islands, with an Index to the Records of the Rarer Species, 8vo, 6s 6d, new 1872

HARTING (J. E.) Bibliotheca Accipitraria, A Catalogue of books, ancient and modern, relating to Falconry with notes glossary, and vocabulary, *illustrated with plates, portraits, coloured fronts*, &c., roy. 8vo, *half mor., cloth sides*, 17s 1891

HARTING (J. E.) The Fauna of the Prybilov Islands, Group of Seal Islands of Alaska *with a plate of 9 figures* 38pp 8vo, 9d 1875

HARTING (J. E.) The Ornithology of Shakespeare critically examined, explained and *illustrated engraved fronts*, 8vo, *cloth* 9s 1871

IBIS. 5th series complete with index 1883—88 in 6 vols, 8vo, *half morocco cloth sides marbled edges*, nice series £5 18s 1883—88

IBIS. 1870 to 1891 complete, 21 vols, 10 vols, *newly bound half brown morocco, top edge gilt*, rest in *cloth* and parts, £26 1870—91

IBIS. 1861 and 1862 complete in 8 parts, clean, uncut as published, scarce, £2 18s 1861—2

JERDON (T. C.) Illustrations of Indian Ornithology, containing 50 *hand coloured plates*, 4to, *new tree calf gilt, gilt edges*, £12 12s 1847

One plate drawn and coloured quite equal to the others.

LAYARD (E. L.) Birds of South Africa, *plate*, 8vo, *half calf*, good copy scarce. 9s Cape Town, 1867

LAYARD (E. L.) Birds of South Africa, New Edition, thoroughly revised and augmented by R. B. Sharpe, *illustrated with most beautifully coloured plates*, thick imp. 8vo, *half morrocco*, £2 8s (pub £3 3s) 1875—84

LEWIN (W.) The Birds of Great Britain, systematically arranged, accurately engraved and painted from nature, with descriptions, 322 *fine coloured drawings of birds and their eggs*, large paper, 8 vols bound in 4, folio, *half morrocco gilt, gilt edges*, £6 6s 1795—1801

A magnificent copy of this valuable work. The illustrations are exceptional in their colouring and the whole work is in splendid condition.

LILFORD (Lord, F. Z. S.) Illustrations of the Birds of the British Islands, *beautifully coloured*, parts 1 to 32 complete, roy. 8vo, 12 *plates in each*.

The drawings are made in the instance by the best artists, and the plates are coloured lithographs, reproduced from these by the best known methods.

LINNEAN SOCIETY'S Journal of Zoology, *illustrated with many steel engravings of insects, mollusca, crustacea, annelida, etc.*, being the Contributions of the most Eminent Naturalists of the present time—Darwin, Wallace, Walker, Blackwell, Baird, Pascoe, Maclaxton, O. P. Cambridge, Bush, Bates, Hincks, Butler, Cobbold, Gray, Lubbock, Halliday, Smith, Westwood, Newport, Lowne, etc., complete from its commencement in 1857 to 1894, 24 vols bound in 20, *half calf, fine set*, £8 8s 1857—1894

——Ditto Another set in parts, 24 vols, 8vo, £6 6s 1857—1894

MARSH (O. Ch.) Odontornithes, a Monograph on the Extinct Toothed Birds of North America, *with* 34 *fine plates and* 40 *woodcuts*, roy. 4to, *cloth*, scarce, £1 10s 1880

MONTAGUE (Col.) Ornithological Dictionary together with the additional Species described by Selby, Yarrell, and in Natural History, Journal, compiled and edited by Edwd. Newman, 8vo, *new cloth extra gilt*, 3s 6d, 3s 10d post free (pub 7s 6d) 1882—89

Only a few copies can be sold at this low price.

MORRIS (B. R.) British Game Birds and Wild Fowl, *illustrated with* 60 *coloured plates*, 4to *half calf gilt*, £2 2s 1855

An original coloured copy.

MORRIS' (Rev. F. O.) A History of British Birds, in 6 vols, super roy. 8vo, *cloth*, *with* 394 *plates coloured by hand*, £4 (pub. £6 6s) 1860—62

———Ditto new edition in parts now publishing parts 1 to 15 now ready at 2s 6d per part.

MORRIS (Rev. F. O.) Natural History of the Nests and and Eggs of British Birds, *with upwards of* 200 *coloured illustrations*, 3 vols, 8vo, *half calf gilt*, First and Best Edition £3 3s 1853

MUIRHEAD (Geo.) Birds of Berwickshire, *illustrated with upwards of* 150 *fine engravings and folding map*, 2 vols, 8vo, *cloth*, as new £1 8s 1889—95

MULSANT (E.) **et VERREAUX** (E.) Histoire Naturelle des Oiseaux Mouches ou Celebres constituant la Famille des Trochilides, *illustrated with* 65 *exquisite coloured plates of birds*, 4 vols, 4to, *new half red morocco, cloth sides, gilt top and uncut edges* together with Supplement, containing 56 *additional superbly coloured plates, in boards*, forming together 5 vols, 4to, *illustrated collectively with* 121 *plates*, £7 17s 6d 1874—77

ORNITHOLOGIA BRITANNICA. Avium Omnium Britannicarum cum Terrestrium, Aquaticarum Catalogus, Sermone, Latina, Anglico et Gallico, redditus, cui, subjicitur appendix avis alienigenas, in angliam raro advenientes, complectens, 1 *plate*, imp. folio, 5s scarce 1771

Printed for the author.

ORNITHOLOGIST (The). A monthly magazine of Ornithology and Oology, edited H. Kirke Swann and others, *illustrated with plates*, 8vo, subscription for the year 6s post free

POPE (A.)—Upland Game Birds and Water Fowl of the United States. 20 *delightfully coloured plates*, equal to drawings, each measuring 22 by 28 inches, and mounted on heavy card boards. 1 vol, atlas folio. *half mor.*, £6 6s

PERSIA, EASTERN. An Account of the Journeys of the Persian Boundary Commission. 1870, '71, '72—Vol 1, The Geography, with Narratives by Major St. John, Lovett and Enan Smith, and an Introduction by Major Gen., Sir F. J. Goldsmid, C.B., *with maps and illustrations*,—Vol 2, The Zoology and Geology by W. T. Blanford, A.R.S.M., *with* 28 *illustrations*, 18 *of them beautifully coloured, and map illustrative of Zoological lines*, 8vo, *cloth*, clean uncut £1 5s (pub 42s) 1876

PETIGREW. On the Mechanical Appliances by which Flight is attained in the Animal Kingdom, 4 *plates*, 4to, 4s 6d *L. S.*, 1868

SAUNDERS (Howard) Illustrated Manual of British Birds, *with beautiful illustrations of nearly every species*, 736 pp, *cloth* quite new 17s 6d 1889

——Ditto new *half morocco gilt* £1 1s

SCLATER. (P. L.) Catalogue of a Collection of American Birds. 20 *beautifully coloured figures in characteristic attitudes* 8vo, *cloth* £1 15s 1862

SEEBOHM. (H.) A History of British Birds, *with coloured illustrations of their Eggs*, this fine work forms 4 vols, 3 *of text and* 1 *of plates*, *cloth*, £5 5s 1883

SEEBOHM. (Henry) The Geographical Distribution of the Family Charadriidæ, or the Plovers, Sandpipers, Snipes, and their Allies. About 500 pages, and 500 *excellent woodcuts by* Messrs. Lodge, Millais, Holding, etc., *cloth*, £1 10s 1888

SHARPE. (R. B.) A Chapter on Birds, Rare British Visitors, *with* 18 *coloured plates, by Keulemans*, post 8vo, cloth 3s 1895

SHARPE. (R. B.) and Wyatt (C. W.) A Monograph of the Hirundinidæ or Family of Swallows, parts 1 to 20 complete, wich 129 *life like plates, exquisitely coloured by hand*, 20 parts in 11, 4to, original wrappers (pub £10 10s) £5 5s 1885—94

SHELLEY. (Capt. G. E., F.Z.S.) A monograph of the Nectariniidæ, or Family of Sun Birds, *with* 121 *beautifully coloured plates*, 4to, *original half red morocco gilt, gilt top edges*, £7 18s London, 1876—1880

A fine, clean, coloured copy of this valuable work.

SWAINSON. (Wm., *F.R.S.*) Birds of Brazil and Mexico, *the illustrations consisting of* 78 *most beauifully, coloured plates drawn to the life equal to drawings*, roy. 8vo *half morocco, marbled edges*, £7 7s very scarce

SWAINSON. (W.) The Naturalists Guide for Collecting and Preserving all Subjects of Natural History and Botany, 72 pp., 2 *plates*, post 8vo, 1s (pub 5s 6d) 1822

SWAYSLAND. (W.) Familiar Wild Birds, complete in 4 Series, *with 40 full-page exquisite coloured illustrations and numerous original wood engravings in each, and descriptive text*, 4 vols, cr. 8vo, *cloth gilt*, £1 10s (pub £2 10s)

THOMPSON. (Wm.) Natural History of Ireland, 4 vols, 8vo, *cloth, nice clean copy, with portrait*, £3 3s 1849—1856

The work consists of the Birds and Mammalia, Vol 1 being the orders Raptores and Insessores; Vol 2, Rasores and Grallatores; Vol 3, Natatores: Vol 4, Mammalia, Reptiles, Fishes, and Invertebrata.

WHEELWRIGHT. (H. W.) Comparative List of the Birds of Scandinavia and Great Britain, sm. 4to, 1s N.D

WHEELWRIGHT (H. W.) A Spring and Summer in Lapland, comprising the Ornithology of Lapland and East Finland, with Descriptions of the Fish, Insects, etc., and Hints to a Bushman, second edition, *with beautiful coloured plates of birds*, cr. 8vo, *cloth extra*, new, 6s (pub 10s 6d) 1871

WHITE. (Rev. Gilbert) Selborne, edited by Thos Bell, *choicely illustrated with plates etc.*, in 2 vols, 8vo, *cloth*, (pub £1 11s 6d) £1 5s 1877

The most complete edition.

WHITE. (Rev. Gilbert). The Works in Natural History of the late; comprising the Natural History of Selborne, the Naturalists Calendar, and Miscellaneous Observations, to which are added a Calendar and Observations by W. MARKWICK *with plates and engravings*, in 2 vols, 8vo, scarce, 15s 1802

WILLIAMS. (T.B.C.) A Bibliography of the Books treating of Fancy Pigeons, with Notes on their rarity and value, 20 pp. 8vo, 6d, scarce, London, 1887

ZOOLOGICAL SOCIETY OF LONDON. Proceedings, complete from 1863 to 1894, with all the *coloured plates*, and general indices, forming 35 vols, 20 bound in *half morocco, cloth sides*, (as good as new) rest in parts, £38 1863—1894

A beautiful series of this useful work.

—— Ditto Another Series *with coloured plates*, 1884 to 1893. 10 vols *cloth, uncut edges*, 8vo, (pub at £24) £15 1884—1893

ZOOLOGICAL SOCIETY OF LONDON. Transactions, complete from vol, 5 to 12 *illustrated with* 669 *plates, many beautifully coloured*, 4to, (pub £97 7s) clean and perfect £18 18s 1862—1890

ZOOLOGIST. A Monthly Journal of Natural History, edited by E. Newman and J. E. Harting, complete from its commencement in 1843 to 1891, in all 49 vols, 8vo, *half calf*, (2 patterns) £12 12s (nearly new) 1843—91

www.ingramcontent.com/pod-product-compliance
Lightning Source LLC
Chambersburg PA
CBHW021942220326
41599CB00013BA/1489

9783743324671